The Plausibility of Life

The Plausibility *of* Life

Resolving Darwin's Dilemma

Marc W. Kirschner and John C. Gerhart

Illustrated by John Norton

Yale University Press New Haven and London

Designed by Sonia Shannon
Set in Bulmer type by Binghamton Valley Composition
Printed in the United States of America by Vail-Ballou Press

Library of Congress Cataloging-in-Publication Data
Kirschner, Marc.
The plausibility of life : Resolving Darwin's Dilemma / Marc W. Kirschner
and John C. Gerhart.
p. cm.
Includes bibliographical references (p.).
ISBN 0-300-10865-6 (cloth : alk. paper)
1. Evolution (Biology) 2. Life—Origin. I. Gerhart, John, 1936- II. Title.
QH366.2.K57 2005
576.8—dc22 2005040113

A catalogue record for this book is available from the
British Library.

The paper in this book meets the guidelines for
permanence and durability of the Committee on Production
Guidelines for Book Longevity of the Council on Library
Resources.

10 9 8 7 6 5 4 3 2

To Phyllis and Marianne

Contents

Preface

This book is about the origins of novelty in evolution. The brain, the eye, and the hand are all anatomical forms that exquisitely serve function. They seem to reveal design. How could they have arisen? The vast diversity of organisms, from bacteria to fungi to plants and animals, all are of different design. How did they originate? Nothing in the inanimate world resembles them. All are novel. And yet novelty implies the creation of something from nothing—it has always defied explanation. When Charles Darwin proposed his theory of evolution by variation and selection, explaining selection was his great achievement. He could not explain variation. This was Darwin's dilemma. He knew only that variation was indispensable as the raw material for selection to act on, and random with respect to the particular selection at work. Genetics provided important clues about the dependence of variation on genetic change and in particular about how change is inherited. What has eluded biologists is arguably the most critical: how can small, random genetic changes be converted into complex useful innovations? This is the central question of this book.

To understand novelty in evolution, we need to understand organisms down to their individual building blocks, down to the workings of their deepest components, for these are what undergo change. Insights into these components have come only in the past few years. A theory of novelty was impossible to devise until the end of the twentieth century; experimental evidence was incomplete on how the organism uses its cellular and molecular mechanisms to build the organism from the egg and to integrate the genetic information into functional processes. Ignorance about novelty is at the heart of skepticism about evolution, and resolving its origins is necessary to complete our understanding of Darwin's theory.

The last 150 years have seen Darwin right and Darwin wrong; Darwin doubted, Darwin ignored; Darwin demonized, and Darwin idolized; but in the end we may have the true worth of his accomplishment. He came up with a single transcendent idea, variation and selection, and he demonstrated that idea through intense observation. This science is the simplest to appreciate; one might even say it is science at its purest. So convinced was Darwin of variation and selection, based on his empirical evidence, that he was willing to ignore or contrive mechanisms to explain it. The course for biologists has been ever more clear: to see if we can understand the mechanistic underpinnings of his transcendent idea.

Evolutionary biologists and paleontologists in their search for more evidence of selection and common descent have done their part, though their task is hardly complete. Geneticists, achieving spectacular success at the end of the twentieth century in solving the mechanism of heredity for all of life, have done their part. Still, they can do more with the modern tools at their disposal.

Developmental biologists, cell biologists, biochemists, and now genomicists have begun the arduous job connecting the bewildering amount of genetic change to the variation on which selection has acted. It is their insights that we report here. An understanding of the connection between the gene, on the one hand, and the anatomy, physiology, and behavior of the organism, on the other, can provide the explanation for novelty. Knowing the ease with which novelty can arise in turn helps us determine whether it is plausible that life is a product of evolutionary change.

In this book we propose a major new scientific theory: facilitated variation that deals with the means of producing useful variation. From an explanation of how such variation emerges comes an appreciation of the facility of evolutionary change. We present facilitated variation not only for the scientist, but also for the interested nonscientist who is ready to explore ideas at the forefront of biological theory. Recognizing how difficult it is to speak to such a diverse audience, we owe both groups an explanation.

To the scientist, we ask forbearance that we have largely skirted the jargon and qualifying phrases emblematic of scientific writing. Yet many of our scientific colleagues who read drafts of this book strongly encouraged us to keep the language simple while making no concessions in the ideas. Even if we had tried to confine the message to professional biologists, we would have had problems. In which subfield would this book be understood? If it were addressed primarily to those who study molecular biology, would the ideas be familiar enough to those who study natural history? If addressed strictly to evolutionary biologists, our assumptions would disenfranchise most molecular biologists, who would find the questions peculiar and the examples exotic. We decided that a common, straightforward vocabulary was essential just to reach *scientists* as a group. To move beyond scientists to the lay public required further adjustments, but fewer than one might expect.

To the nonscientist, we would say that you have already revealed your deep interest in evolution and your appreciation that evolution affects your sense of self as a biological creature. In record numbers you have bought books, visited museums, traveled to exotic habitats, and attended courses and debates about evolutionary theory. Your intense demand for knowledge has been met by interpreters of science, often journalists, who have contributed to your understanding. But the barrier of ignorance of the molecular sciences has handicapped the lay public, as it has in fact handicapped many scientists as well. To be forced to occupy the worst seats in the theater for one of the most meaningful dramas in the history of human exploration seems tragic, especially if it is avoidable. The nineteenth-century discoveries in evolution filled museums with towering fossil skeletons of dinosaurs, which inspired children and adults alike. Zoos, arboretums, and animal programs on television have thrilled millions with the diversity of life on earth.

We are not sure that we can succeed as well in portraying the molecular and cellular understandings that complement and ultimately explain this diversity. But we know from experience that a vivid real drama can be much more engaging than a paraphrased retelling. We have done what we could: reduced the jargon, emphasized the universal concepts, stayed true to the narrative of evolutionary history, and provided a glossary and ongoing explanations. What we have not done is dilute the ideas or turn arguments and demonstrations into uncorroborated assertions. We have tried to provide conveniences and aids, but there is no shortcut to understanding. We hope we have succeeded in both explicating a significant new theory in evolution and embracing a broad audience.

As an original, far-reaching recasting of evolutionary theory, our book has much to convey. We have high drama: the union of molecular, cellular, and developmental biology with evolutionary history; the story of how novelty was generated in evolution; the paradox of the conservation of fundamental mechanisms of the cell but the extraordinary diversity of organisms; a new cast of evolutionary mechanisms all based on trading constraint for deconstraint; and the completion of Darwin's theory with new evidence as to why his original idea of variation and selection works on the variation side as well as on the selection side. We hope that the magnitude of a retold story of creation will hold the interest of readers—specialists and generalists alike.

Ours is a journey from molecule to cell to organism to life's diversity. It is up to the reader to traverse the nearly four billion years of life embedded in our account. We have invoked the latest results from the molecular sciences, pressing chemistry, cell biology, developmental biology, biochemistry, and genetics into the service of evolutionary biology.

Understanding life is not a conquest, but a slow lesson in appreciation. Most of what we, the authors, have learned we learned from others; our own contributions are small enough that they rarely appear in this book. We, as scientists, have been and continue to be active participants in the process of discovering how the organism constructs itself. We continually confront the surprising admixture of conservation and diversity found in all organisms. Our lifelong pursuits of the conserved processes of life led us inexorably to the question of the origin of novelty in evolution. Novelty by definition is always a surprise, but when the surprise is too great, it is completely implausible. The plausibility of life rests on the plausibility of generating novelty, and that in turn rests on mechanisms newly uncovered in biology.

We thank all those who read the manuscript in its entirety and provided suggestions for improvement: Spyros Artavanis-Tsakonas, Jean Thomson Black, Walter Fontana, Peter Gray, Saori Haigo, Daniel Kirschner, Elliot Kirschner, Donald Lamm, Richard Lewontin, Christopher Lowe, Charles Murtaugh, Clifford Tabin, David Wake, Rebecca Ward, and Mary Jane West-Eberhard. We are very grateful to Donald Lamm for his steady encouragement and wise suggestions throughout this project and to Jean Thomson Black for her literary advice, and to Vivian Wheeler for her careful editing. We appreciate the elegant artistry of John Norton and the continuing administrative support of Yolanda Villarreal Bauer. Finally, we thank Phyllis Kirschner and Marianne Gerhart for creating the environment in which all of this could happen.

The Plausibility of Life

INTRODUCTION

A Clock on the Heath

In 1802 the Reverend William Paley expressed his faith that life, full as it is of intricate design, must be the work of a Supreme and Intelligent Creator. In his now-famous metaphor, the minister wanders on the heath and stumbles across a brass watch. Plunged into thought, he asks how the watch came into being and reflects that his explanations are entirely different from those brought to mind when his boot hit a stone. The stone might have "lain there for-ever," demanding no explanation. But the watch, with its carefully constructed wheels, teeth, springs, pointers, and oval glass face, each part perfectly suitable for the function of telling time, certainly must have been created by a designer of great skill. Even if the watch were broken or if we did not understand the workings of every part, our confidence in the existence of a designer would not be shaken. No one, Paley asserts, could believe that a purely blind and random process of trial and error could achieve the exquisite design of the "plainest" parts of the watch.[1]

Paley intended his homily to demonstrate the need for a Creator in life's creation. "Every observation . . . concerning the watch may be repeated with strict propriety concerning the eye, concerning animals, concerning plants, indeed all the organized parts of the works of nature." These works are far grander than a mere watch. As human beings are the only designers capable of creating a watch but are incapable of creating life itself, it is fair to deduce that a far greater Intelligent Creator of life must exist or must have existed.[2]

Paley compared the complexity of the watch, which he could understand, with the complexity of life, which in 1802 he could not, as a measure of their creators. However, such comparisons look different today. Where he would have seen an earthworm and a skylark each as a unique and complex design, we now see underlying similarities; they have the same system of heredity, the same genetic code, the same cellular makeup, the same subcellular components, largely the same metabolism, and many of the same processes of embryonic development. Paley was on a firm footing in distinguishing the stone and the watch, but not in comparing the watch and the skylark, the worm, or the eye. He had every reason to see each as an independent act of creation. All he saw in common was their complexity, not the nature of the complexity, and it is that nature that tips the balance between acceptance of evolution and the alternative deism that Paley chose.

Fifty years later Charles Darwin guessed right. In the 1850s only a little more was known about the constituents of living things, such as the existence and continuity of cells. Darwin used his imagination to replace a supreme designer with a process of evolution by natural causes. He theorized that in a population of organisms, minuscule heritable variations of design arise at random in each generation, and some rare variant members are by chance more fit to reproduce under the selective conditions, a process known as survival of the fittest. As the other designs are rejected, the altered design of the survivors is perpetuated. Evolutionary adaptation is improved design for life.

Here and throughout this book we use the word *design* to mean a structure as it is related to function, not necessarily implying either a human or a divine designer; it is a commonly used term in biology. With time, according to Darwin, large novelties of design accrued from sequentially selected small novelties. As the process was repeated (and as the lineage of descendants repeatedly branched), a single primordial cell gave rise to all life forms on earth, including human beings. It might take a long time with many individuals dying in the line of service, but better adaptations would eventually result from the modification of previous adaptations, toward the same or new purposes.[3]

Neither Paley nor Darwin could directly observe the events of creation. Both Divine Creation and evolution by variation and selection were hypotheses. In the 150 years since Darwin, natural selection has been amply demonstrated by biologists who have trolled the ocean and scoured forests and barren lands to identify new species and unearth fossils. But does natural selection fully explain the diverse complexity of life on earth? Darwin himself waffled about the relative importance of variation and selection for the creation of novelty. Was variation rare and channeled in specific favorable directions? Or was variation so common that any trait would be likely to occur at some frequency?

Initially, Darwin thought that variation was common and therefore selection was for him the only creative force in evolution. Variation was required, but selection molded the chaotic profusion of small changes into the exquisite design of organisms. In this light, variation seemed less important than selection. In later life, though, Darwin gave variation a larger role in evolution, though not a freely creative one. He accepted the view that the environment directly instructs the organism how to vary, and he proposed a mechanism for inheriting those changes. He retreated from the notion that variation was random with respect to environmental conditions. The more important he made the environment in determining the kind of variation, the less was its importance as a selective and creative agent.

This ad hoc theory was at first proclaimed as Darwin's second monumental achievement, after the theory of evolution. Yet it was completely wrong. The intuition that served Darwin the naturalist so well in the *Origin of Species* failed him when he tried to understand cellular mechanisms and inheritance. In the years after Darwin, his original ideas were restored. Variation was again seen as random and providing the essential material on which selection could act. Variation was recognized as the source of novelty; the environment could not produce anything new through the selective process.

The notion of random variation as the sole generative force behind novelty raised other problems as well. Darwin worried about complex organs such as the eye, where multiple independent events must have preceded the appearance of the first working eye. An eye requires a lens to form images, and a retina of photoreceptors to receive them, and long nerves to communicate signals from the retina to special parts of the brain. Would the intermediate eye be any more functional than a partially assembled watch? If not, how were intermediates maintained so that slowly over time new parts could be added until a selectable function was achieved? Though anticipation and planning to meet multiple demands are common tasks for intelligent beings, they are hard to achieve by random variation and selection.

Thus, the problem of novelty's origin in evolution becomes, How could the eye be created in the first place, or the brain, or wing, or lungs, or limbs? Could they have been plausibly assembled, small piece by small piece, each presupposing a selective advantage? It is this feature of Darwin's theory, the uncertain accounting for novelty, that creationists seize on; meanwhile, evolutionary biologists assert that variation must be sufficient, though they lack a general explanation for the origin of complex novel structures. Answers to these questions affect the plausibility of life's arising by way of evolution.

Science in Darwin's time could not provide satisfactory answers about the nature of variation. Darwin simply chose a catechism different from Paley's on which to base his interpretation of creation, namely, that heritable variation is generated by some means, and selection then sifts the variants for those most reproductively fit. It was an interpretation that we now recognize as modern, completely based on natural events and laws, but one that better describes improvements than it does origins. It gives us no idea of how fast or how readily things could change, or whether evolution is channeled in certain directions by the kind of variation that an organism can produce. To this day, the explanation for novelty has remained hidden within the organism. Paley went straight to an ultimate cause: a Creator about whose means of creation we can know nothing more.

For a while in the twentieth century, the concept of the gene and mutation seemed to provide the answer to evolutionary change; namely, if a gene is altered by mutation, the descendants inherit the change, and depending on the nature of that particular change, the descendant would differ in some trait of its anatomy, physiology, or behavior. It now appears that the concept provided only a partial answer, that genetic change is required for heritable variation. Genetics tells us a great deal about the inheritance of change and the spread of the required gene in a population of reproducing animals when the trait is under selection. Still, it does not tell us much about how genetic change causes complex changes in organisms. Only in the last few decades have such cellular and developmental mechanisms been identified. These mechanisms speak most directly to the question of the origins of novelty.

To show the vantage point of our times, let us imagine a twenty-first-century descendant of Paley, more than two hundred years removed from the author of the homily on the watch, wandering the heath and still wondering about the origin of plants and animals. She brings with her an education in modern biology, including genetics, cell and developmental biology, and evolution. She does not have the good fortune to stumble upon a brass watch (they are getting harder and harder to find), but instead muses philosophically about life itself, the heather, the flies on the heather, or the mouse underfoot.

Like her famous ancestor, she is fascinated by measuring time. She notices that plants extend their stems below the flower just before sunrise. She notices on a longer time scale that some plants flower early in the season when days are short, whereas others flower at the peak of the summer when days are long. She notices in herself that she has a daily cycle of sleep and restlessness and that she has suffered recently from jet lag, thereby raising her personal awareness of her endogenous clock. She realizes that most kinds of plants and animals, even fungi and bacteria, have such clocks; being experimentally inclined, she might have placed a mouse in total darkness and found that its 24-hour cycle of sleep and waking continues for many days without cues of light. As an avid student of time, she might know that accurate time pieces were once difficult to make, especially ones that kept time when jarred or heated or cooled. By comparison, biological clocks function accurately in animals while they run, jump, and swim through life, on hot or cold days.

Also, Paley's modern descendant is understandably impressed to know that virtually every cell in our body, each weighing less than a billionth of an ounce, contains an accurate temperature-compensated chronometer, whereas the first accurate chronometer in human history, circa 1736, weighed 72 pounds (33 kg).[4]

By now the younger Paley, seeing the performance of biological clocks, might be even more tempted than her ancestor to invoke the Creator. But living in the twenty-first century, and with her background in modern biology, she can examine for herself the workings of the biological clock in a way her forebearer could not. She avails herself of the electron microscope, the various tools of molecular biology, the geneticists' collections of mutant animals and plants defective in various aspects of their timing, the sequences of the genomes of numerous animals and plants, and the computerized databases available worldwide.

On her worktable she quickly assembles the clocks of human beings, mice, flies, fungi, and plants. These are known as circadian clocks from the Latin *circa*, approximately, and *dies*, day. How are they constructed? Are they fashioned out of special materials, unknowable to humans? Do they work by means beyond her comprehension? Is each a unique event of creation, different from all other circadian clocks? Does their design offer clues about the designer? Does each clock so far exceed human imagination in its uniqueness, complexity, and perfection that it could never have arisen by the gradual modification of parts affected randomly by mutation and then selected? Or might there be a surprise here, an unexpected glimpse of a plausible creation by natural means?

Man-made clocks, like biological clocks, run by converting a continuous process into a repetitive process. Although they share this common principle, their inner workings are distinctly different. The Chinese water clock of the eleventh century was based on the periodic filling and emptying of vessels attached to the rim of a wheel, into which water flowed at a constant rate. The pendulum of a grandfather clock is kept in motion by weak nudges from falling weights. The oscillating escapement of a brass watch is driven by an uncoiling spring.

The quartz watch uses an electrical current to cause a crystal to vibrate at a characteristic frequency. Though all convert a continuous process into a periodic one, they share few components of their internal time-keeping mechanism.[5]

Unlike man-made clocks, circadian clocks from disparate sources share many features of design and materials. Turning to the components of the clock, the modern Paley would find that most are used elsewhere in the organism in other roles having nothing to do with clocks and are far from being unique. They are all made of proteins and most of these proteins resemble other kinds of proteins. Furthermore, when she compares the components of the circadian clock in the fruit fly with those in the mouse, she finds that many of them are the same, but some are used differently in the two circuits. The interactions of the different clock components are not strictly conserved, but they can still generate periodic behavior. It is as if the genes and encoded proteins act as individual transistors suitable for wiring in different ways in the integrated circuit timers of a mouse or of a fly.

Thus, the circadian clock is not like a brass watch, where each component is made for just one purpose. The human-engineered clocks use different techniques to achieve the same result; the circadian clocks use a common set of techniques. Novelty in human clocks requires independent acts of invention. Novelty in biological clocks seems more suited to iterative modification from a common origin.

No matter where she turned, whether to the nervous system, the embryo, or the behavior of cells, young Paley would find examples of multiple and varied reuse of the same components. The properties of components facilitate their reuse, new use, and rampant invention. She would not find a boundless variety of completely different objects performing complicated activities, of the sort that demand a supreme Intelligent Designer to explain their origin. She would not even be tempted to follow the trail in that direction, so enthralled would she be by what organisms have managed to do with the limited cellular components at hand.

Indeed, a similar moment of introspection arose for many biologists in the year 2000, with the publication of the "rough draft" sequence of the entire human genome. It was realized that we possess 22,500 genes, only six times the number possessed by a bacterial cell, the simplest of all known free-living organisms. How could human complexity be achieved with so few genes? Then, in the next few years, the genomes of bacteria, fungi, plants, fish, and mice were sequenced and compared, and it turned out that many genes are similar across these disparate species, apparently conserved from remote ancestors. How can their differences of anatomy, physiology, and behavior be explained when many of their genes are so similar?

The answer, the young Paley infers, lies in the multiple use of versatile conserved components. It is not the clock in particular that is so remarkable, but the multifunctioning protein components and their forms of regulation that allow them to be easily connected in many ways toward various ends. The living organism is certainly more complex than the brass watch in terms of the number of components and the variety of their interactions, but it is complex in unusual ways appropriate for versatility and modification rather than for dedicated single use. In the end, the young Paley would conclude that biological clocks do not imply a human creator or a divine Creator, but something else—call it a creation of biological novelty through natural causes.

Our story of the two wanderers on the heath brings us to the heart of this book. We begin where the younger Paley left off, at the question of the origin of complex life. We bring to the inquiry the understanding of many processes of living organisms, not just clocks, gained in the past few decades by a worldwide community of biologists. It is an understanding obtained at the level of the chemical components of organisms, their activities, and their interactions, with glimpses of their evolution.

The cardinal issue in evolution is the origin of complex and heritable variation from a limited reservoir of components. Although selection has preoccupied evolutionary biologists, the study of the origin of variation and novelty has idled. Is the organism's capacity to generate heritable variation great enough to supply the succession of variants needed for natural selection to bring forth a circadian clock, or—more challenging—a human being from a single-celled ancestor, all within the time span of the earth? Heritable variation requires mutational change of the genome, but that is only the start of the story.

What else is required to get an adequate frequency of selectable variants? Mutation only changes what already exists. It does not create new anatomy, physiology, and behavior from nothing, so we need to know how readily one structure can be transmuted into another, particularly when we consider structures of intricate design and interdependent activities. With an understanding of how random genetic change is converted into useful innovation, a theory of novelty can be devised. Darwin's general theory of evolution can then be established at the most fundamental level.

ONE

The Sources of Variation

Physical scientists in the nineteenth and early twentieth centuries had astounding success in formulating very general yet predictive theories in thermodynamics, electricity and magnetism, and atomic structure. Biology sought a similar level of generalization and had signal success in the cell theory, the germ theory of disease, metabolism, and heredity. Darwin's theory of evolution was perhaps the most ambitious effort to understand the living world, but unlike the others it was historically based and hard to test experimentally. Even to be comprehensible, it required an accumulation of knowledge from natural history, genetics, and paleontology. Biology differed from physics in that its most obvious characteristics are complexity and diversity; therefore the origin of that complexity and diversity would remain at the center of biological concerns. At the end of the nineteenth century, evolution was an unfinished, still controversial theory. An explanation of the origin of variation was one of the big gaps. The incompleteness of the theory was a problem for all of biology; biologists would continually return to it to add their perspectives.

The Three Pillars of Darwin's Theory of Evolution

Darwin's all-encompassing theory of evolution was based on three major supports: a theory of natural selection, a theory of heredity, and a theory of the generation of variation in the organism. In Darwin's view, rephrased in modern terms, organisms within populations vary genetically and consequently differ in traits that affect their capacity to contribute to the next generation. In competition with one another and facing other pressures in the environment, the most fit organisms flourish and the less fit fail to contribute progeny to the next generation. This process selects a better-adapted subset of the population that carries within it a different set of genes and therefore manifests a different set of properties. The population is said to have evolved under selection, making use of its genetic variation.

From the start, it was natural selection or the struggle for existence that required the smallest leap of imagination. The selective death of "weaker" individuals is universally appreciated; Darwin employed his encyclopedic knowledge of biology and his persuasive logic to draw out the consequences in a robust argument that has lasted to the present day. Artificial selection was familiar to all plant and animal breeders, and the extrapolation over long periods to a "natural selection" was plausible. Still, some critics denied its effectiveness as the only mechanism for producing very large anatomical changes, such as had occurred in the evolution of complex animals from single cells or of human beings from animals.

By contrast, heredity was not properly understood in Darwin's time. Today it is largely a solved problem. With the deciphering of the structure of DNA in 1953 came a sophisticated understanding of genetic variation and its inheritance. Genetic variation is due to mutation (a change in sequence of the chemical letters that make up the DNA code—A, T, G, and C), to recombination (the splicing together of DNA segments from different chromosomes to form hybrid chromosomes), and to assortment of chromosomes during egg and sperm formation. All of these factors, separately and together, change the DNA sequences of an offspring, and these changes are reliably inherited.

For some biologists, basking in the grand accomplishments of genetic theory, understanding that DNA changes its sequence randomly at very low frequency (a few positions in a billion bases of sequence, each round of replication) and that DNA is otherwise copied at high fidelity at each cell division, meant that a comprehensive theory of heredity could join a well-developed theory of selection to complete Darwin's transformative idea. For others, a major weakness remained, casting all else in doubt. Their unanswered question was whether random change and shuffling of DNA could ever lead to highly complex and wonderfully adaptive innovations in anatomy and physiology such as the eye, the brain, or even the peacock's tail. The Reverend Mr. Paley's skepticism, shared by some scientists as well as by many laypeople, might not be satisfied by a theory of evolution that rested solely on a theory of selection and a theory of the inheritance of random DNA changes.

In evolution, selection always acts on variation of the *phenotype*, which includes all the observable and functional features of the organism. This is a favorite word of evolutionary biologists, as in "phenotypic variation" or "phenotypic change." Selection does not directly act on the DNA sequence (also called the *genotype*). It acts on the genotype only indirectly through the phenotype, most details of which depend on the genotype. The organism's size, its speed, its visual acuity, its resistance to disease, its behavioral responses—all are part of the phenotype. DNA itself has none of these activities. Since the phenotype faces selection but the genotype is what is inherited to produce the phenotype, it is crucial to understand the processes that connect the two.

The question unanswered by the two well-established pillars of evolutionary theory (selection and heredity) is whether, given the rate and nature of changes in the DNA, *enough of the right kind* of phenotypic variation will occur to allow selection to do its work, powering complex evolutionary change. If the organism were a machine, like Paley's watch, we would expect that random alterations either would have little effect or would lead to catastrophic failure. We would not expect random change to cause the clock to run more accurately or to develop new features, such as a snooze alarm! But is an organism like a watch, or is it made in a fundamentally different way? The question was unanswerable until the very end of the twentieth century. No clues existed in Paley's or Darwin's time. In later chapters we will argue that understanding the organization, growth, and development of the organism is essential to complete Darwin's theory.

There are limits on what selection can accomplish. We must remember that it merely acts as a sieve, preserving some variants and rejecting others; it does not create variation. If genetic change were random, what could ensure that enough favorable phenotypic variation had taken place for selection to have produced the exquisite adaptations and variety we see on the earth today? At various times, biologists thought that genetic change must be directed in some way to produce enough of the appropriate kinds of phenotypic variation. If selection were presented with a preselected subset of variants, that might greatly facilitate evolutionary change. Or if the organism generated just the right variants, selection might not be needed at all. Thus, the efficacy of selection would depend on the nature of phenotypic variation, which in turn depends on the amount and type of genetic variation and on the mysterious process by which phenotype emerges from genotype.

Is genetic variation purely random, or is it in fact biased to *facilitate* evolutionary change? By *facilitated genetic variation,* we mean genetic variation that would be (1) biased to be viable (only nonlethal variation is heritable, the rest from the point of view of evolution is useless); (2) biased to give functional outcomes; and (3) biased to be relevant to the environmental conditions. A few biologists tried to invent theories about how the environment might alter the parents' genetic endowment to their offspring. As attractive as it would be to discover a process for loading the genetic dice, thereby improving the rate and course of evolution, there is in fact no evidence for facilitated *genetic* variation and there is conclusive evidence that it does not exist. The process of evolution receives no help from this quarter, and within our modern understanding of the organism it would be hard to imagine how such a process could work.

By 1940 it was clear that genetic variation was random and unlinked to environmental conditions. Stripped of these concerns, evolutionary biologists formulated a theory based on purely random (unfacilitated) genetic variation and on selection. This was the Modern Synthesis, the current consensus model of evolution. The theory, however, was codified before the dawning of modern molecular biology, cell biology, and developmental biology (a more modern term for embryology, which includes the study of stages other than embryos). Evolutionary biologists could not say much, even in theoretical terms, about how the organism constructs itself, its phenotype, from its genetic instructions, its genotype. The Modern Synthesis is a valuable model but an incomplete one. It lacks the third pillar required of a general theory of evolution, a pillar needed to explain the feasibility of evolutionary change.

This third pillar is a theory of how genetic variation is used in the generation of heritable phenotypic variation. It is a theory of how the inherited genetic material along with the environment constructs the individual organism in each generation, from the egg to the adult and on to the next generation. The organism's anatomy, physiology, and behavior are only remotely connected to the DNA sequence through all the complex processes of growth, development, and metabolism, though they depend on it. A change in the DNA sequence is therefore only indirectly correlated with a change in the anatomy and physiology of the organism.

Currently, our understanding of this connection is not sufficient for us to predict the phenotypic consequence of most genetic changes. We can identify genes that predispose a person to cancer, but we cannot draw a perfect correlation between gene and disease. Given the remote connection between the DNA and the phenotype, we have no way of knowing how often random DNA modification can produce useful outcomes for selection. Without an understanding of how DNA changes are interpreted, we cannot know how much selection molds evolution, or how much the initial variation biases the outcome.

It is not enough to know that changes in DNA can in some unknown way cause a change in phenotype; we need to know at least in outline how phenotypes respond to particular changes in DNA. It is this third pillar, an understanding of the organism's response to genetic change, that is our subject here and the resolution to Darwin's dilemma. The overall outline of this very modern story is so new that it is only dimly perceived, and its implications for evolution have been only partially discerned. Before we consider the role of the organism in responding to genetic change, we need to understand whether there is an environmental bias in how the DNA changes. Then we can add the final pillar to Darwin's basic outline and construct a more plausible and more complete theory of evolution.

How Random is Variation?

Among the first ideas of how variation might be generated and inherited were those of Jean-Baptiste Lamarck (1744–1829), a French biologist who was among a group of scientists believing, as Aristotle had, that organisms have changed over time rather than having been fixed since the moment of their creation. Like Paley and many others, Lamarck had marveled at how well living organisms are adapted to their environment. He looked for a means to match these adaptations to environmental conditions during the course of evolution. In his *Philosophie zoologique*, published in 1809 (the year of Darwin's birth), Lamarck proposed two laws. The first restates a common observation about use and disuse: "In every animal . . . more frequent and sustained use of an organ strengthens that organ . . . while the constant disuse of an organ imperceptibly weakens it . . . and ends in its disappearance."[1]

His second law is novel and extends the process of adaptation to the generation of heritable change: "Everything that nature has caused individuals to acquire or lose by the influence of the circumstances, it preserves by heredity and passes on to the new individuals descended from it."[2]

In Lamarck's view, an animal's perception of and response to stressful circumstances is based on physiological and behavioral needs, not emotional and conscious desires. He focused on the influence of behavior on evolution as a stimulus for evolutionary change.

Lamarck's best known example is the giraffe. He supposed that the pre-giraffe, in meeting its need to feed, stretched its neck and forelegs. The human neck can adapt physiologically, as illustrated in Figure 1. In Lamarck's view, an acquired physiological adaptation of a

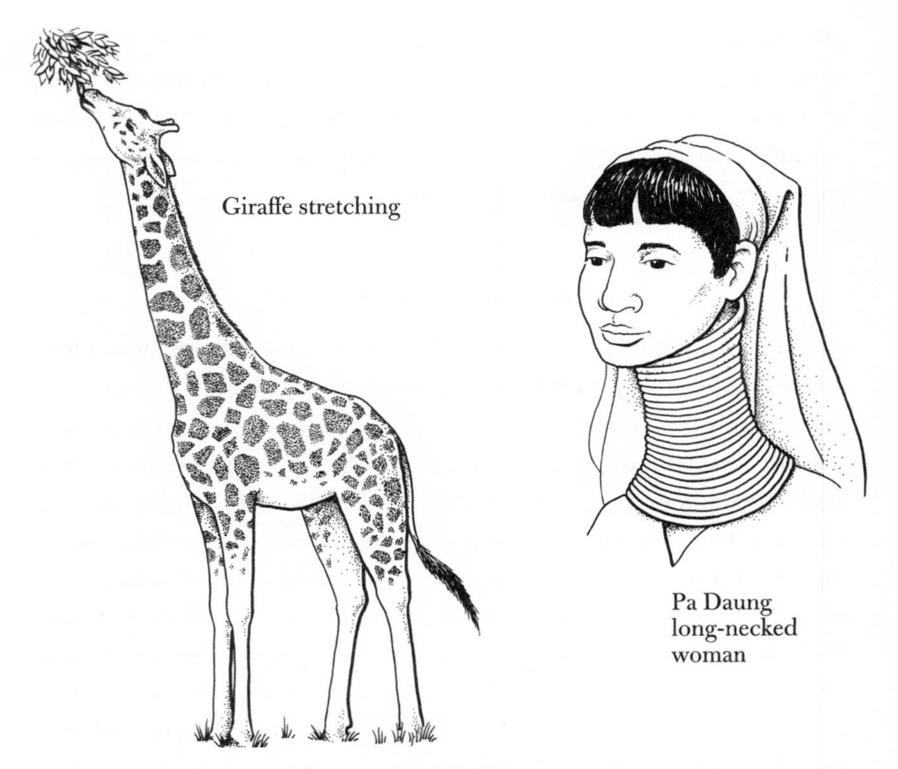

Figure 1 The elastic neck and Lamarck's theory of evolution. *Left*, the giraffe stretches its neck for food. *Right*, a woman of Pa Daung, Myanmar, with neck rings. In neither case is increased neck length, gained by stretching, transmitted genetically to the next generation.

longer neck and forelegs in the giraffe (Figure 1) was passed to the offspring who continued stretching until the long-necked, long-legged giraffe of many generations later did not need to stretch any farther. Presumably, many members of the pre-giraffe population could change as a group, not just as a rare giraffe variant.

Another example for Lamarck was the pre-ibis or pre-crane, which realized the need to keep its feathers dry, stretched its legs to rise above the water, then lengthened its bill to reach the fish in the water, and stretched its toes to create large webbed feet—accomplishing all this over many generations. Presumably, many members of the population changed together. It was a perfect gradualist idea. Throughout this transformation, the behavioral need drove the anatomy. "It is not the shape either of the body or its parts which gives rise to the habits of animals and their mode of life; but it is, on the contrary, the habits, mode of life and all the other influences of the environment which have in the course of time built up the shape of the body and of the parts of animals."[3]

So self-evident and appealing seemed the view of facilitated heritable change, or more commonly called *inheritance of acquired characteristics*, that Darwin himself could not escape using it. Even though he proposed in his 1859 *Origin of Species* that change was random and selected later, he felt his hypothesis incomplete until he could identify how heritable variation arises in the first place. He increasingly retreated to Lamarck's view that different circumstances evoke different responses in organisms, which somehow pass to the next generation; that is, the environment facilitates or induces the kinds of adaptations appropriate to the environment. In 1868 Darwin published his two-volume work on *The Variation of Animals and Plants under Domestication*, presenting his model of the inheritance of acquired characteristics. In justifying his surrender to an overtly Lamarckian theory, he wrote:

> How again can we explain the inherited effects of the use or disuse of particular organs? The domestic duck flies less and walks more than the wild duck, and its limb bones have become diminished and increased in a corresponding manner in comparison to those of the wild duck. A horse is trained to certain paces, and the colt inherits similar consensual movements . . . How can the use or disuse of a particular limb or part of the brain affect a small aggregate of reproductive cells, seated in a distant part of the body, in such a manner that the being developed from these cells inherits the characteristics of one or both of the parents? Even an imperfect answer to this question would be satisfactory.[4]

Darwin "imperfect answer" was pangenesis, his theory of inheritance. In pangenesis, the parent's entire body influences the next generation by influencing the germ cells (egg and sperm). In this way, novelty is more efficiently generated in the offspring. Darwin suggested that minute elemental particles—we would now call them informational particles—are given off by all cells of the body and circulate through the individual. The more a cell is used, the more particles it gives off. Eventually the particles concentrate in the germ cells, their numbers reflecting the adult's lifetime of experience and physiological adaptation to the environment. Once passing from the germ cells into the embryo, they affect the development of the offspring by emphasizing those aspects that had been most called upon in the previous generation. The representation of these elemental particles in the germ cells seemed to reflect actual physiological usage rather than perceived needs, so the idea was not as need driven as Lamarck's. Variation was not random, but directed by circumstances, and was carried into the offspring by the sperm or the egg.

Darwin's was a self-consistent theory for the generation of variation, based solely on the inheritance of acquired characteristics. Pangenesis was an immediate success, but its author had created a dilemma for himself. The more successfully an animal generates appropriate variation in response to the local environment, the less the local environment needs to act through natural selection, preserving one variant from a multitude of others. In the extreme, there would be no need for natural selection at all; the organism would merely change as dictated by the environment. Darwin seemed to conflate variation and selection, and this fusion demanded further explanation.

In retrospect, it was difficult for thinkers of the time to break from the notion of directed heritable variation (although Darwin himself had done so earlier in his first theory) and to accept the possibility of pure random variation. It was hard to imagine that random events on their own could create a kind of novelty adaptive to the selective conditions. By direct reference to the ultimate physiological target, pangenesis or any of the other non-Mendelian ideas of directed inheritance avoided the need for stepping-stones to the new phenotype. Attractive as these ideas were, they were completely without foundation.

The Disproof of Facilitated Genetic Variation

Subsequent years of experimentation brought no support for the direct inheritance of physiological adaptations of the organism to the environment. On the contrary, substantial evidence accumulated for the view that pangenesis did not exist.

The first steps in distinguishing physiological adaptation (the subject of Lamarck's first postulate) from heritable variation (the subject of his second postulate) came in 1895 from August Weismann. He showed that it was extremely unlikely that the sperm and egg could receive any information from the environment.

Weismann asked a simple anatomical question, "Where in the developing embryo are the specific cells that later become the eggs and sperm in the adult?" Studying jellyfish, he found that germ cells of the adult arose from precursor cells that were clearly segregated from other cells. Only after the jellyfish developed to the adult stage did these cells migrate into the gonad from their isolated site.

The initial segregation of the germ cells has been confirmed by modern studies of many kinds of animals including insects and all vertebrates, as illustrated in Figure 2. The cells of the body, called somatic cells (from *soma*, the Greek word for body), are the ones that experience and respond to stresses of the environment; they make no contribution to the distant germ cells, which are the only cells able to contribute to the next generation. The germ cells, by comparison, are shielded and removed from environmental influences. Weismann considered Darwin's pangenesis theory to be completely ad hoc. He criticized it delicately: "His [Darwin's] assumptions do not, properly speaking, explain the phenomena. They are to a certain extent a mere paraphrase of the facts . . . based on speculative assumptions."[5]

Weismann's idea of the soma–germ-line distinction has stood the test of time. No influence from the external environment that impinges solely on the somatic cells can modify the hereditary material for the next generation, which is exclusively within cells of the germ line. Conversely, since germ cells do not perform physiological functions in the organism and hence cannot be directly selected upon by the ex-

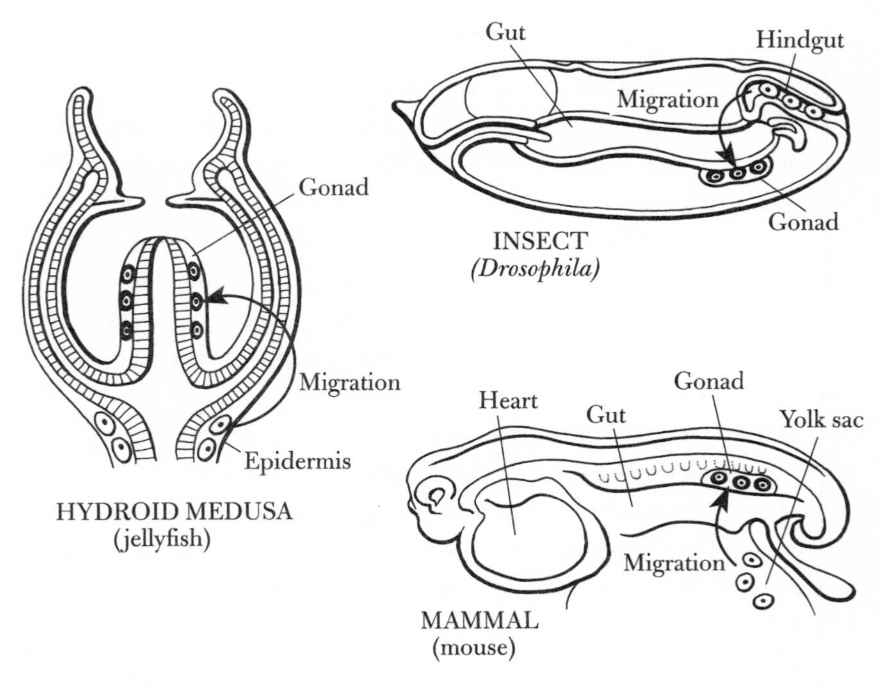

Figure 2 The separation of the germ line and the body. Germ cells are initially separate from cells in the developing embryo; later they migrate into the gonads of the embryo and differentiate into eggs and sperm. *Left*, the jellyfish example discovered by August Weismann. *Upper right*, an insect. *Lower right*, a mammal.

ternal environment, nothing in the environment can influence them to transmit specific traits preferentially to the next generation. The germ cells are coselected as mute passengers, in the vehicle of an organism made up of somatic cells with the same genetic makeup as they have. The only selection that can take place is the survival and reproductive success of the entire individual derived from a fertilized egg. A principal biological advantage for sequestering germ cells from the soma may be to assure that they reflect only the success of the whole individual, instead of the success of any selfish somatic lineage of cells within the individual that could most influence them. Weismann added to Darwinian evolution the cell-biological evidence that nullified the concept of the inheritance of acquired characteristics.

Looking for Macromutations

Although Weismann produced strong arguments against facilitated genetic change, they did not quite sound the death knell for that idea. Variation and heredity were certainly the heart of evolution, and an understanding of the nature of variation in clear chemical and physical terms would be necessary before one could claim that the facilitation of variation could not occur by biasing genetic change. Hence, understanding variation and settling the issue of whether it was random or not was extremely important. As we shall see, the problem of genetic transmission became such a compelling problem in its own right that it quickly eclipsed the problem of evolution.

Before the twentieth century's rediscovery of Gregor Mendel's work, the importance but not the nature of variation was evident. The future geneticist William Bateson wrote in 1894: "Variation, whatever may be its cause, . . . is the essential phenomenon of Evolution. Variation, in fact is Evolution. The readiest way then, of solving the problem of Evolution is to study the facts of Variation." He scoured the world for freaks of nature—human feet with eight toes, turtles with two heads, horses with an apparent atavistic formation of multiple metacarpal bones, insects with all limbs duplicated—enough to convince himself that variant individuals occur in populations at detectable levels.[6]

Bateson was famous for his study of "homeotic" variation (meaning a change into the likeness of something else), which resulted in a class of variants with serial repetitions of anatomical features such as extra digits or wings. The common appearance of well-proportioned duplicated appendages might have been a clue that variation was distinctly nonrandom. Later, homeotic variation experimentally induced by mutation would provide a critical insight into how the organism develops. These insights, in turn, provided a key to our theory of facilitated variation.

However, in 1894 Bateson must have been very disappointed. In his nearly six-hundred-page book entitled *Materials for the Study of Variation*, he could provide no mechanism for the origin of homeosis or any other kind of variation, except to say that embryonic development had been altered.

Although the study of phenotypic variation was proving to be a dead end at the turn of the twentieth century, the study of heredity was bursting with opportunity. The second pillar of Darwin's theory was about to be triumphantly established, but ignorance of the mechanism of phenotypic variation has lingered until the present. At first an understanding of genetic variation, coupled with a theory of selection, seemed all-powerful; but the problem that was actually solved was how information was transmitted from one generation to the next, not how novelty originated.

The modern story of genetics began in 1900 with the rediscovery of Gregor Mendel's 1866 paper "Experiments in Plant Hybridization." By then, many biologists and naturalists were breeding plants systematically and observing the distribution of the phenotypic differences to the offspring. Thus, Mendel's paper could be resurrected and appreciated. Variation in organisms could be divided, as we have said, into two categories: genetic change and phenotypic change. Genetic change occurred in the abstract but increasingly manipulatable realm of the organism's genotype (now defined as the information coded in the genome, the DNA sequence of the four chemical letters A, T, G, and C), whereas phenotypic change took place in the observable but still baffling realm of the organism's anatomy, physiology, development, and behavior—some of which was heritable and some of which was dictated by the environment.

After the successful sequencing of many genomes of bacteria, fungi, plants, and animals, information about the genotype is in principle both precise and complete. Plainly stated, the genotype is the organism's DNA sequence. There is no ambiguity. In the early days of genetics, the genotype could only be inferred from mating experiments, using some element of the phenotype as an indicator of the state of the genotype. (Mendel used color and texture of peas, for example.) The use of the phenotype to signify the genotype was an unavoidable but indirect method. Today the genotype can simply be read out as a sequence from the DNA, over a billion letters long for many animals, much like a computer program.

The phenotype is a much more daunting matter. To understand it means to understand all the events of embryonic development, growth, maturation, and experience of the organism. It is everything that contributes to what an organism is, what it looks like, how it functions, and how it behaves. It is easy to see why, before Mendel's work was rediscovered, phenotypic variation was the major question for people interested in evolution; it was what was observable and what natural selection directly acted upon. After 1900, however, understanding of genotypic variation and the transmission of genes became major concerns for geneticists. Phenotypic variation was put aside. Bateson and others turned to the new field of genetics (he invented the word) as a full-time pursuit, with increasingly less regard for the problems of evolution.

Understanding evolution had been very much in the minds of the early geneticists; it was often the question that impelled them to enter science. Thomas Hunt Morgan, who later became indisputably the greatest American geneticist, visited the garden of Hugo de Vries in Holland in 1900 to examine for himself the first evidence for *macromutation*, the supposed evolutionary transformation of one species into another in a single mutational event.

What Morgan would have seen in de Vries's garden was an assortment of aberrant evening primroses, collected by de Vries, that had arisen occasionally and spontaneously in neighboring fields. The aberrations were drastic, creating what seemed like new species within a generation or two. Some plants had red veins on the leaves instead of colorless veins, some were larger or smaller, some possessed smoother or longer leaves, and some had modified flowers, as illustrated in Figure 3.[7]

Macromutation seemed to solve many of the problems of heredity and evolution in Darwin's theory of gradual change. Small changes would not need to accumulate over many generations, with each generation running a very real risk of being diluted by interbreeding with

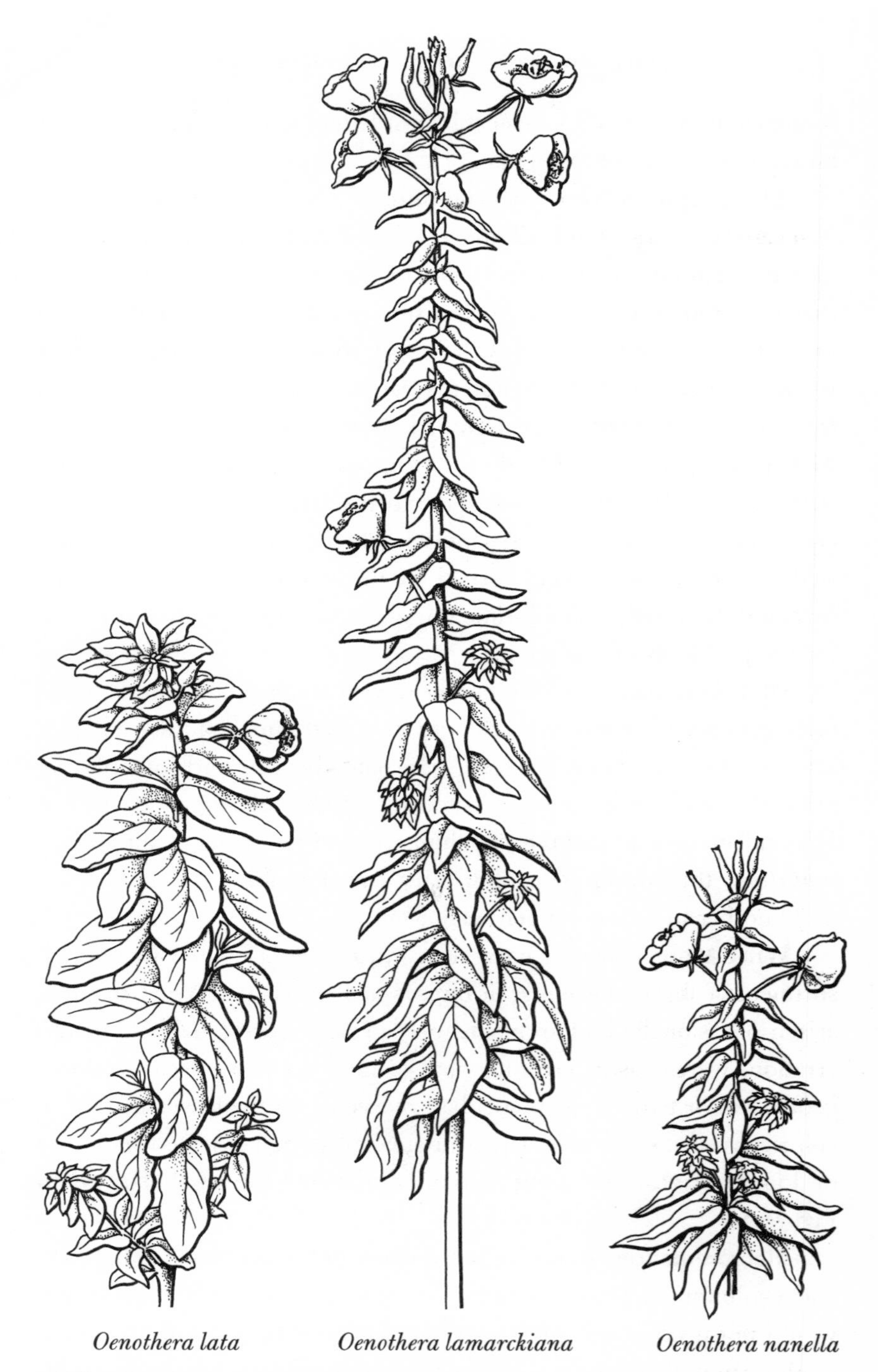

Figure 3 Macromutations in the evening primrose, *Oenothera*. *Center*, the stock cultivated by Hugo de Vries. *Left and right*, two short "species" that suddenly mutated from the stock.

normal individuals. The mutant plants, because of their large differences, might prevent interbreeding; since they arose commonly enough, those of similar kind could isolate themselves into new breeding associations. The sudden appearance of a new species could explain the gaps in the fossil record.

As a convert to the experimental method, Morgan ultimately sought ways to test de Vries's theory in other organisms. He would fail completely. We now know that the macromutation in the evening primrose is not a general phenomenon but is caused by a rare and peculiar genetic mechanism of that hybrid species. Macromutation in the evening primrose was a complete dead end for the study of evolution.

Morgan chose the fruit fly as a subject to test the generality of de Vries's observations. He propagated fruit flies in the dark for many generations to see if their eyes diminished, perhaps in a single stroke, like the macromutations in the evening primrose. There was no loss of eye structures even after 49 generations in the dark, therefore no evidence for hereditary loss through disuse either rapid or slow. Morgan generated many small heritable changes of phenotype, which were viable and fertile, but found no massive transformations like the evening primrose macromutations.

Then, one day in 1910, Morgan found a peculiar mutant fly that changed the course of biology. It was not a dramatic mutation, but was unusual. It marked Morgan's transition from an experimental evolutionary biologist to an experimental geneticist. Concurrently, interest in evolution was generally declining among many mainstream biologists.[8]

T. H. Morgan's mutant fly was a white-eyed male found in a population of normal red-eyed flies. Morgan had found a gene for eye color on what we now know as the X chromosome, a sex chromosome present in males in only a single copy, but in females in two copies. Using this mutation, he proved by various matings of mutant and normal flies that chromosomes determine sex, which had already been indicated by observing chromosomes through the microscope. This discovery marked the beginning of a demonstration of many of the basic and universal facts of genetics. Morgan and his students introduced genetic mapping as a means to establish the order of genes on the chromosome; it is the basic technique used today to map the genes of human disease, such as Huntington's disease and cystic fibrosis.

Discoveries in genetics typically used inbred strains of animals, plants, and fungi with the particular species chosen for their tolerance to life in the laboratory. Morgan initially turned to inbred strains because animals from the wild, when mated, produced offspring with too much variation in their traits, such as wing size or eye color. But with this choice he turned from wild populations where the dynamics and variation of populations could be observed, to inbred laboratory strains where they could not. Variation, previously a source of fascination, was becoming an experimental nuisance. Selection was now performed in the laboratory by geneticists to identify traits that were easy to score, rather than traits that might be related to survival in the wild, or to embryonic development, or to evolution. The original impulse to understand how organisms evolved was lost.

Morgan and his group helped to initiate the modern field of experimental genetics. They found no evidence that genetic variation was directed; all their data were consistent with random genetic change. As a footnote, Morgan personally maintained his broad interest in developmental biology and evolution until his death in 1945. By 1928, when he moved from Columbia University to the California Institute of Technology, he gave up work on the fruit fly and turned again to issues of variation and individuality. None of his famous students followed this path.

The Last Hurrah for Facilitated Genetic Variation?

By the dawn of molecular biology in the 1950s, there had been no credible evidence that an organism could specifically respond to an environmental stress by mutating a particular gene. But a well-designed molecular experiment disproving such a Lamarckian connection was lacking. To resolve this question once and for all, John Cairns, a well-known bacterial geneticist and biochemist, turned to the human gut bacterium, *Escherichia coli*, which can be studied in large populations and over many generations.

Cairns asked if bacteria repaired a specific damaged gene at a higher rate when it was needed for growth than when it was not needed. Could the bacterium generate a heritable response to need, in the same way that giraffes extended their necks when they *needed* food that was out of reach? To the surprise of the scientific community, Cairns at first claimed evidence for directed genetic (Lamarckian) change. The bacterium repaired the gene by mutation (reversing or compensating for the initial DNA sequence change by further changes) at an accelerated rate if it needed the enzyme for growth, faced only with the alternative of starvation and eventual death.[9]

Though the gene was repaired, the critical question was whether it was modified more quickly than other genes that were not required for growth. With further analysis, it turned out that Cairns had badly misinterpreted his result: stressful starvation conditions increased the mutation rate for *all* genes, not just for the required gene. This increase of rate was an adaptation to the stress of starvation, which subsided when the bacterium grew again. Subtle technical reasons had completely fooled Cairns; once again, the search for Lamarckian inheritance had failed, here under conditions that many biologists considered the most favorable for finding it, if it existed.

After half a century of molecular biology, we still have found no mechanism that, as a physiological response of the mother or father to environmental stresses, modifies the genetic information of the egg or sperm. We know that various viruses can carry genetic information into cells, and this information can be incorporated into the cells' DNA, their permanent genetic dowry. The prevalence of viral sequences in the genome suggests that viruses at various times have entered the germ line from the outside. Again, there is no evidence that the genes carried by these viruses reflect any previous stress-related physiological response by the host.

Modern molecular and genetic analysis has revealed no hint of directed genetic change in response to physiological need or experience. No mechanism is known to direct a specific environmental stress toward the alteration of a specific gene or set of genes, as a way to ameliorate that stress. Hence there is no evidence for "facilitated genotypic change." Genetic variation and selection are completely uncoupled.

The Modern Synthesis

By 1940 the liberation of genetic change from Lamarckian overtones (even without Cairns's negative experiment) allowed leading evolutionary biologists to come together and proclaim a modern Darwinian theory, the Modern Synthesis. Close to Darwin in all important respects, it was now made consistent with contemporary science. Competing theories of evolution rapidly lost favor. The macromutations of de Vries had sunk to the status of a special case. Orthogenesis, a view that organisms evolve according to internally directed rather than externally selected paths, was simply a misinterpretation of the variation in existing populations. It was not so much false as it failed to offer any mechanism.

New traditions emerged with the study of wild populations, which drew on entirely Darwinian concepts. Population genetics resolved some problems that Mendelian genetics seemed to have created for traits that were continuous and quantitative rather than discrete, as well as explaining how a genetic change spreads in a population. Natural selection was given center place in sifting "profligate and chaotic" variation into the diversity of organismal forms we know.[10]

Stephen J. Gould argued that the Modern Synthesis had quickly hardened into a strictly adaptationist program, focusing on selective conditions and ignoring the role of the organism in generating phenotypic variation. By 1940 the fossil record had grown, and more and more gaps seemed to be filled. Although the record was known to be incomplete, it was surprisingly compatible with Darwin's ideas. Several specimens of *Archaeopteryx* had been found (the first in 1861), and its partial reptile–partial bird traits seemed to imply a smooth progression toward birds as we know them, not a macromutational eruption of birds from reptiles.

Since 1940 at least 12 other relatives of intermediates of the reptile-bird lineage (feathered dinosaurs, miniraptors, and flying dinosaurs) have been unearthed—many in China in the 1990s, though none is quite like *Archaeopteryx*. A recently discovered example of a feathered dinosaur is shown in Figure 4. The assemblage of fossils suggests ordered changes in the feathers, reversal of the pubis bone in the pelvic region, reversal of the first toe, and reduction of vertebrae in the tail.

Yet in this worldview that saw all creativity in evolution as coming from selection, something was missing. It is as if a play had been written, the stage was set, but the cast had been forgotten. The organism and its role in creating variation were largely absent.[11]

The Modern Synthesis of 1940 was not so much wrong as it was incomplete. Biology itself had deeply split, perhaps making completion more difficult. The three great disciplines—genetics, developmental biology, and evolutionary biology—had gone separate ways. When new fields such as molecular biology and cell biology emerged, they had essentially no contact with evolutionary biology.[12]

The Mendelian understanding of heritable variation was the principal advance of the Modern Synthesis. Heritable variation was divided into two parts: variation in the genotype and variation in the phenotype. After making the important distinction that only the genotype is inherited but only the phenotype is selected, the Modern Synthesis reduced evolution to three basic steps. First, there was the occurrence of random genotypic variation—in modern parlance, a random modification of the sequence of DNA. Second, the change of genotype caused a change of phenotype within the individual organism (by means not specified). Third, the altered phenotype was selected (and with it the altered genotype required for it) on the basis of the individual's reproductive fitness, that is, its ability to contribute progeny to future generations.

On the question, of how the altered genotype caused an altered phenotype, the Modern Synthesis was silent. The old ideas of the environmental induction of variation had been purged. A key tenet of the synthesis was the independence of phenotypic variation from ambient selective conditions. Experience, learned behavior, or physiolog-

Figure 4 A feathered dinosaur: *Protarchaeopteryx* reconstructed from 125-million-year-old fossils from northeast China. Its length was 2 feet (70 cm). Having feather-like outgrowths from the integument, it is considered a member of a flightless dinosaur group sharing a common ancestor with birds. (Redrawn from Angela Milner, "Dino-Birds," Natural History Museum, London, 2002.)

ical adaptation to the environment could not be inherited. No explanation was offered to replace environmentally induced variation. The evolutionary biologists of the mid-twentieth century cannot be faulted for failing to explain variation, for only the first ingredients of an explanation were yet available. A molecular theory of genetics was 15 to 20 years away, and a molecular theory of comparative embryology would only come at the very end of the twentieth century. Instead, biologists might be faulted for their failure to recognize this large gap in their evolutionary theory. They mostly just ignored it.

Despite this omission, evolutionary biologists maintained strong views on the nature of phenotypic variation. Many thought that anything in the phenotype could change owing to random mutation. According to Gould, Darwin thought that variation must meet "three crucial requirements: copious in extent, small in range of departure from the mean, and isotropic" (or undirected toward adaptive needs of the organism). Gould called these three attributes of variation Darwin's most brilliant insight, "because he realized that selection could not otherwise operate as the creative force in the evolution of novelties." The alternative would have the organism generate a biased profusion of phenotypic variation for selection to act upon.[13]

The Modern Synthesis made the concept of adaptation paramount in evolutionary theory. The organism was like modeling clay, and remolding of the clay meant that each of the billions of little grains was free to move a little bit in any direction to generate a new form. This was close to saying that not only was the input of genotypic variation random but the output of phenotypic variation was random as well, or at least constrained very little. With this approach, the problem of how the processes of embryonic development and cellular function create the phenotype could be largely dismissed as interesting but not informative for evolutionary change, further segregating evolutionary biology from its peer disciplines. Selection alone might suffice to understand the succession of phenotypes that constitutes the history of evolutionary change. If an organism needed a wing, an opposable thumb, longer legs, webbed feet, or placental development, any of these would emerge under the proper selective conditions, with time.

The organism, it seemed, could be counted on to generate all of the variation needed for selection to act.

Some biologists later argued that the organism was constrained as to the kind of variation it could produce: rather than the full panoply of changes, some kinds would be missing. Perhaps some components of the organism were more difficult to change than others, and these would remain unchanged. Indeed, many conserved proteins and genes exist in the phenotype. In general, though, constraint was considered a minor effect, or trivial, for example, in explaining why mollusks and echinoderms were less able to evolve wings than vertebrates.

Novelty, Time, and Random Mutation

What if evolutionary biologists were wrong to think of phenotypic variation as random and unconstrained, even though genetic variation was random and unconstrained? How much would it matter if we really understood how genetic variation leads to phenotypic variation, and in particular how facile or difficult is it to achieve a specific phenotype? Well, we could perhaps say we understand how evolutionary change occurs, based on the organism's capacity to generate novelty, without reference to particular selective conditions or catastrophic events. Also, we would be able to face the issue of the rate of evolution, which has always been imponderable. Skeptics of evolution, even in Darwin's time, said that the hypothesis of selection acting on variation certainly sounds reasonable, but there has not been enough time for suitable variants to arise. Organisms just could not generate bat wings and whale flippers by variation and selection in the twenty million years indicated by the fossil record. Shades of Paley's argument about the watch!

By comparison, if we question how long it would take a high-speed computer to write randomly a specific Shakespearean sonnet, we are asking that all the letters of the words of the sonnet will come up simultaneously in the correct order. It is an impossible task, even if all the computers in the world today had been working from the time of the big bang to the present. Even to compose the phrase, "To be or not to be," letter by letter, would take a typical computer millions of years.

Of course, the chance of coming up with a specific sentence or sonnet would improve vastly if selection or biased variation were introduced. On the selective side, we might accept provisionally a partial success such as "Tu is or no to iz" and then improve upon it—but that would already be lowering our selective requirements. Or we might keep individual correct letters as they arise, rather than waiting for all the correct ones to come up at once. Biasing variation can also improve the rate of outcome: if the computer generates only known words (using a dictionary) rather than random letter combinations, the process is accelerated. And if the computer generates only English words of three or fewer letters, the time to get the sentence is shortened to much less than a year. Thus, biasing variation should also have a huge effect on the speed of evolutionary change. Finally, if biased variation and piecemeal selection are combined, the required time can be very short.

Many evolutionary biologists dismiss the issue of rates of variation. They tell us that geological time is, in fact, very long when compared with the decades, centuries, or millennia that have sufficed for the divergence of domestic animals into grossly different breeds by artificial selection, or for the changed coloration of moths or beak size of finches via natural selection. Admitting all this, some skeptics are still not willing to grant that random variation can produce anything as complex as a flower or an eye, even over geological time, much less a human being from a bacterium-like organism.

Without some account of how complex novelty arises, mere refuge in the sufficiency of time is unconvincing. To comprehend fully how genotypic change generates phenotypic change, one needs an understanding of how the genotype generates the phenotype. A degree of understanding is coming where none was before, giving us a sense of the ultimate map between genotype and phenotype. That map should provide a way to estimate the feasibility of evolutionary change. The existing phenotype of the organism biases the realm of possible phenotypic variation: that much is self-evident. But how, how much, and in what directions it biases novelty in evolution remain difficult and crucial questions.

Toward a Theory of Facilitated *Phenotypic* Variation

As we have seen in this chapter, genetic variation is not channeled toward adaptation to selective conditions. Whatever bias there is to alter the amount and kind of phenotypic variation must arise out of the construction of the organism itself. Our theory about how organisms generate novelty in evolution starts with some assumptions which, though not in dispute, are not commonly appreciated.

First, genetic variation is required for evolutionary change. Genetic variation initially arises by mutation. Much of the genetic change that is important in evolution comes from the reassortment of mutations of previous generations by sexual reproduction.

Second, present-day organisms come from previous organisms, so they may retain remnants of the properties of their ancestors, including properties that allowed them to change in the past. A big surprise of modern biology has been conservation—that even distantly related organisms use similar processes for cellular function, development, and metabolism. Each process, comprised of many protein components working together, contributes to the phenotype. When a process is conserved, most of its protein components are conserved. Details of metabolism are the same in bacteria and humans; basic cell organization and function are similar between yeast and humans; and developmental strategies in fruit flies are strikingly similar to those in humans. The conservation of key processes in diverse organisms today implies, as we shall see, that we can deduce the basic physiological and developmental processes of organisms in the past. Even though these processes are not revealed by the fossil record, broad conservation among living organisms puts us in an unambiguous position to extrapolate back to our ancestors.

Third, all organisms are a mixture of conserved and nonconserved processes (said otherwise, of unchanging and changing processes), rather than a uniform collection of processes that change equally in the course of evolution. Novelty in the organism's physiology, anatomy, or behavior arises mostly by the use of conserved processes in new combinations, at different times, and in different places and amounts, rather than by the invention of new processes.

We have not yet described the processes themselves, but we shall see that they can be used in many different contexts and to different degrees. This versatility, part of the remarkable adaptability of processes to conditions, is key to their special role in evolution. The surprisingly small number of genes for humans and other complex animal forms reflects the anatomical and physiological complexity that can be achieved by the reuse of gene products. The conserved processes are fundamentally cellular processes; they operate on many levels in the development and functioning of the organism. They are the *core processes* of the organism.

Central to our argument is that these processes, many of which have been conserved for hundreds of millions or even billions of years, have very special characteristics that facilitate evolutionary change. They have been conserved, we suggest, not merely because change in them would be lethal (although that might be a factor), but because they have repeatedly facilitated changes of certain kinds around them.

Many of the conserved core processes have the capacity to be easily linked together in new combinations. New linkages can occur with a minimum requirement for genetic change and hence can happen readily. A new combination of processes can arise with little or no change of the units themselves. We will talk later about the concept of weak regulatory linkage, which means essentially that links between processes can be forged without extensive retooling of each component. To maintain these links, processes are often reinforced with additional weak linkages—the suspender and belt approach to reliability.

Until we describe specific mechanisms, a metaphor may be useful. To double the size of Paley's brass pocket watch, virtually every component would have to be retooled, from the glass face to the brass gears. If growth of an animal involved such a process, it would be nearly impossible. The components of living things are more like Lego blocks. Size and shape of the organism or an anatomical part of the organism can be varied by reusing common components in new combinations and amounts. The blocks do not change but their arrangement does. Linkages are readily made and broken.

In conjunction with the unchanging aspects of phenotype, we have also asked what really does change on a cellular and molecular level in evolution. It is not the conserved core processes. We argue that regulatory components are the main targets for heritable change—small features of the protein, RNA, or DNA that determine the time, circumstances, and degree of activity of the processes. These are often involved in controlling the linkage and activity of processes. Although the phenotype may play out at the gross anatomical and physiological level, the real locus of change is in the cellular processes that generate theses anatomies and physiologies. Sewall Wright, the great population geneticist, said it most clearly: "The older writers on evolution were often staggered by the seeming necessity of accounting for the evolution of fine details . . . , for example, the fine structure of all the bones . . . Structure is never inherited as such, but merely types of adaptive cell behavior which lead to particular types of structure under particular conditions."[14] It is remarkable that in 1931 Wright could foresee a time when it would be possible to explain anatomy and physiology in terms of the cell's adaptive responses to differing conditions. We will show that such adaptability is built into most of the cell's conserved core processes.

Why would organisms be constructed to facilitate evolutionary change? What is in it for them? There are several answers, but the most powerful is that organisms are always changing and responding to change. In the course of life, they alter their physiological state and behavior. They have mechanisms to resist extremes of temperature, to adapt to variations in the food and water supply, and to modify their response to predators. Some kinds of adaptability operate on a short time scale, such as the fight-or-flight response involving adrenalin secretion in threatening situations. Rapid changes occur in the heart rate, vascular system, and nervous system to mobilize reserves. The frightened organism is in a very different physiological state than the secure one. Other kinds of adaptability operate over longer times, such as the acclimation of an animal to high altitude, and even longer-term adaptations of muscle and bone growth in response to repeated exercise or physical load. Physiological adaptability toward environmental change helps the organism survive.

Furthermore, adaptability of the organism is perhaps even more extensive toward the changing internal conditions wrought by embryonic development. Most of this developmental adaptability is invisible to us, because it is directed internally as one group of cells responds to signals from another. However, there are examples of developmental adaptability toward external conditions as well. Although physiological and developmental adaptations operate differently than evolutionary adaptations, they often entail the same cellular mechanisms. The road to evolutionary change is paved with physiological adaptability. Phenotypic variation, and along with it evolutionary change, is facilitated by simple regulatory tweaks to existing physiological and developmental processes that long ago were designed so that the organism could adapt to its environment.

TWO

Conserved Cells, Divergent Organisms

Sewall Wright, who with J. B. S. Haldane and R. A. Fisher established population genetics in the early twentieth century, asserted that beneath the anatomical changes in evolution are changes in what he called adaptive cell behavior. He thereby alleviated some of the difficulties of imagining the evolution of complex organisms. Yet his assertion avoids the obvious next question: What is "adaptive cell behavior"? We now know that the cell has hundreds of behaviors or activities that involve conserved core processes. When a significant change occurs in evolution, do radically new behaviors develop or does the cell use its existing repertoire in different ways?

To understand evolutionary change in Wright's terms, we want to know the historical changes that have occurred in cell behaviors and trace these modifications to the large-scale changes in anatomy that have traditionally been used to document evolutionary history. Yet information about alterations in cell behavior cannot be derived from the fossil record. It is only available from interrogating and comparing extant forms of life. If it were to turn out that cellular change in evolution was idiosyncratic and spread throughout all processes of the organism, it would be difficult to construct an interpretable history of life in terms of changes in adaptive cellular processes. Changes on the cellular level would be chaotic and confusing. Each evolutionary change would be a unique perturbation of the cell's vast program of responses.

What seemed like a potentially insurmountable problem has been rendered feasible by finding that there is only a limited, though large, set of core cell behaviors, which change in limited and understandable ways. Novelty usually comes about by the deployment of existing cell behaviors in new combinations and to new extents, rather than in their drastic modification or the invention of completely new ones. True novelty in the invention of cellular processes is rare. Once such novelty occurs, it may be carried through stably in many lineages. Hence, evolution is divided into epochs of invention of cellular behaviors, interspersed with long periods without invention. During these extended periods a lot is happening, though. The novel deployment of conserved cellular behaviors continually gives rise to new phenotypes.

In this chapter we identify some of the rare moments when new cell behaviors were invented, and we look at their emplacement. Rather than analyzing the conventional progression of anatomical forms, we provide a narrative history of the invention and use of these core processes in evolution. Because of their stability over long periods, we call these cell behaviors *conserved core processes*. Adaptability is one of their characteristic traits. We will provide evidence for the view that evolutionary changes in the anatomy and physiology of organisms involve the reuse of these conserved core processes in new circumstances. Reuse itself implies that the core processes are constructed in such a way as to facilitate phenotypic variation.

Evolution from the Perspective of the Cell

Modern organisms and fossils constitute our only way of extracting the evolutionary history of anatomical novelty. Fossils present a map of time and structure with many gaps and many uncertainties, but remain our only record of past anatomical inventions. Molecular biologists have added considerably to that archive with a large body of information obtained from sequencing the DNA of extant forms and a few recent fossils to establish the relatedness of organisms by descent.

The similarity of sequences in related forms provides a new type of evidence for deducing the ancient lines of descent through common ancestors, much like a family tree.

In addition to providing their DNA sequences, modern animals give us critical comparative information about the processes of embryonic development, which in turn provide opportunities for understanding how complex anatomical features must have arisen. Because we have no knowledge of the DNA sequence of ancient forms, we triangulate our limited information from extant forms to infer genetic changes in the past, and we map those onto the proposed branching pathways of descent. It is not unlike a crime scene investigation, with partial information and probabilities attached to various scenarios. However, unlike criminal investigations, only one puzzle needs to be solved, and mountains of new information continue to accumulate. Whereas all of this information might have led us into a morass of inconsistency, instead it has gradually reinforced a consistent pattern of descent back to our earliest ancestors. While not underestimating some ambiguities, we are and should be excited by the overall consistency.

The record of living forms also shows a spectrum of anatomies and a parallel spectrum of gene sequences. For example, on a DNA sequence level and on an anatomical level, we are indisputably very closely related to chimpanzees, more distantly related to birds, and even more distantly to fish. But sequence comparisons can lead us beyond anatomical comparisons. Because DNA sequences can be related across all organisms, they can be used to trace descent into the

remote past—even to when we emerged from a bacterial lineage, where no anatomical signposts exist.

The progression of cellular mechanisms is another extremely important feature that we can extract from the study of extant species. We project DNA sequences onto the historical map of descent derived from comparative morphology. From the studies of gene sequences we can tell, for example, when innovations in metabolism occurred, when new sensory modalities arose, and when the complex immune system was created. Even though genes change in sequence over time, we can usually recognize them by sequence features alone, especially using sophisticated pattern-detection computer programs. All mammals, but only mammals, contain the genes for hair proteins, whereas birds contain the genes for feather proteins. All vertebrates contain genes for cartilage; and all animals, but not bacteria, contain the genes for proteins to wrap DNA into chromosomes. Just as anatomical features have a specific time of appearance in history, so must the genes that underlie these features.

Often genes appear long before their modern function is fully realized. Consider, for example, some of the proteins involved in milk production. Found in modern reptiles and birds, they must have existed in the reptilian ancestors before mammals separated from them. Presumably the earlier proteins had different purposes.

Genes for specific structures (hair, cartilage, feathers, milk proteins) pale in significance compared to genes that have been involved in the major inventions of evolution, such as those that supported the first complex cells or the first multicellular animals. No trace of these genes exists in fossils, but the genomes of the millions of extant species, when compared to one another, contain a dense record of past modification. This fact was of only hypothetical interest in evolutionary biology until the molecular revolution. With the advent of huge depositories of information from genomic sequence, it now seems possible to reconstruct the molecular ancestry of extinct forms from the gene sequences of living forms.

Evolution from the Perspective of DNA

Comparisons of DNA base sequences have produced a flood of more objective information about relationships between organisms that is generally consistent with the branching pattern of descent derived from comparative anatomy and the fossil record. As in any new method, serious errors can be made by extrapolating with limited data, but the data themselves are clear and easily verifiable. All living beings from bacteria to fungi to plants and animals have DNA as their genetic material, differing in the orderings of the four nucleic acid bases, A,

T, G, and C. Computerized methods can be used to estimate the minimum number of steps required to evolve one sequence from another.

From these methods we can derive family trees, called phylogenies, where we can relate organisms that look completely different, such as fungi and animals. This approach works because some DNA sequences are very similar, though not identical. For example, when we compare the sequence of a specific gene common to two organisms, of the thousand bases in the gene nine hundred might be identical, with enough similarity to relate them yet enough difference to distinguish them. (One has to take into account that some of the apparent identity is the result of a second mutation; if A is converted to G and then G is later converted back to A, the analysis will score it as no change at all.) Even brewer's yeast, which is a fungus, has sequences similar to those of humans.

It is indisputable that these genes did not move recently from one organism to another but that the identical regions have been maintained for billions of years. Generally, the more sequence differences there are between two organisms, the longer ago was their last common ancestor. Thus, one can construct a "tree of life" containing all modern life forms, all descended from a common ancestor that existed perhaps three billion years ago. Such sequence-based trees can be made without knowing what protein the sequence encodes, if any, or what function has been preserved. The organisms are nonetheless related by descent through shared ancestors.

Reconstruction of the pathway of descent from DNA sequences may be readily appreciated from an example involving reconstruction of the original lost manuscript of *The Canterbury Tales*. The tales were not published in Chaucer's lifetime (circa 1343–1400), and changes and errors must have successively accumulated in the manuscript as it was recopied in the fifteenth century. Researchers used the algorithms of DNA comparison to reconstruct the lineage of the copies. The English alphabet is different from the four-letter DNA alphabet, but the principles are the same—mistakes in spelling and word order are usually propagated. Similarly, the order of the tales corresponds formally to the order of the genes on the genome. Once the gene order is changed, it is unlikely that it will be put back exactly as it was. The results show that a small number of texts served as sources for later texts. It is a tree with several roots. It appears that the earliest manuscripts were different from one another, suggesting that Chaucer probably wrote several copies and distributed them to a few people who assembled them differently.[1]

A similar analysis of Boccaccio's *Decameron*, a collection of a hundred tales written at about the same time (circa 1350) suggests that Boccaccio distributed a finished manuscript. Like the tree of life, and unlike *The Canterbury Tales*, the *Decameron* has a single root and many branches.

Text tracing, whether in literature or in the field of genomics (the study of gene sequences), tells us little about the motivation for the changes. Were some completely accidental? Were some neutral, simply alternative spellings of a word that did not change the meaning? Were others made because the scribe thought he was improving the manuscript? That would be a form of selection. Since we can read the words in *The Canterbury Tales* and interpret their meaning, we can make educated guesses as to what is accidental and what is purposeful.

When it is a matter of the changes of A, T, G, and C in the genome, we cannot easily distinguish meaningless changes from meaningful ones—at least not yet. Words or gene sequences mean little outside the context of their use. Most DNA is presently uninterpretable, and some is undoubtedly of no use at all. However, we can trace the relationships of descent even without knowing the meaning. What we cannot do yet is interpret the consequences of the sequence changes for the organism and its phenotype.

Many of the biochemical pathways in distantly related organisms (such as humans and bacteria) are nearly identical in the chemical transformations they accomplish. Likewise, the functional components of these pathways, the enzymes, are similar in sequence in different organisms. As rapid DNA sequencing became possible, molecular biologists began to see a nearly universal pattern: similar functions were carried out by proteins that had extensive similarity in amino acid sequence, encoded by genes of similar DNA base sequence. This association between similarity in function and similarity in DNA sequence has held up even in distantly related organisms. Therefore, these pathways are preserved much as they must have been in an ancient ancestor.

It need not have been this way. Function might have been conserved and the components changed, or vice versa. Instead, function and protein structure were conserved together in the core processes. Thus, evolutionary pathways may be deployed in different circumstances, but often the pathways themselves are conserved down to both the structure of the circuit and the composition of the components.

Could sequence similarities arise by convergence, that is, by random variation and selection from different starting points? That possibility is statistically highly improbable, because hundreds or even thousands of positions in the DNA sequences are identical. Convergence is the demon of all reconstructions of relatedness among organisms. There are unambiguous cases of its occurrence in anatomical features. The familiar Northern Hemisphere mole, which is a placental mammal, resembles the Australian mole, which is a marsupial. Both have evolved similar adaptations for burrowing and for subterranean life but from different starting points.

In some instances the convergence is more subtle, really a parallelism. The bat wing and the bird wing are anatomically different enough so that there is no doubt that the bat wing is not a direct modification of the bird wing. However, both are modifications of the basic vertebrate forelimb. So there is convergence in function but divergence from the ancestral process of forming a limb.

The comparisons of gene sequences have resolved issues of convergence, even across widely separate lines. The genes are similar by

conservation from a common ancestor, not convergence from different ancestors. As molecular biologists inspected the genes for various functions in bacteria, they found that more than half of them are extensively conserved with genes of human beings, even though their anatomy, physiology, and behavior could not be more different. What

does this mean? Did these conserved components have nothing to do with anatomical and physiological evolution since they did not change? If so, what other parts of the genome did change and were responsible for the change of anatomy and physiology during evolution?

The History of the World According to Genes

The fossil record is a chronicle of both gradual anatomical innovation and gaps that can be interpreted either as rapid change or as deficiencies in the record. What is less apparent is whether molecular and cellular changes occur rapidly or slowly, in large or small steps. The question of step size in evolution has been contemplated since Darwin, who thought that evolution proceeded in small steps, gradually molded by selective conditions. Niles Eldredge and Stephen J. Gould introduced the term *punctuated equilibrium* in 1972, to describe the uneven rate of change of the anatomy of organisms in the fossil record. In their view, long periods of stasis were punctuated by bursts of innovation.[2] We elect to preserve their evocative term but to show that its most important and convincing use is on the molecular and cellular level.

The history of cellular innovation is a story of both conservation and diversification. Core processes have been introduced at rare intervals during evolution (the punctuated part), then are largely unchanged until the present (an equilibrium or stasis). In contrast to the Eldredge and Gould view of punctuated equilibrium, during the periods of stasis in the core processes we see no stasis in anatomical and physiological phenotypic variation in the animal kingdom, which continues undiminished. This theory is more big bang than punctuated equilibrium; once a class of innovations arises, it is permanently retained without further change.

The rest of this chapter is an unusual history of evolution: not a history of the emergence of different anatomies, but a history of the emergence of "adaptive cell behaviors," which are thereafter conserved. When evolution is described in terms of both conservation and diversification of organisms, we get a distinctly different impression from the standard views of evolution, which rely on differences alone. The analysis of DNA allows us to identify the major innovations in the past three billion years. This ancestry is shown in Figure 5, where the innovations and conservation in cell organization are juxtaposed to geological events. There are various lineages we could follow, such as that to plants or to protists or to unusual microorganisms, but largely because of species chauvinism we will follow the trail primarily toward humans.

The following history demonstrates that conservation and diversification are intermingled in all organisms. Conservation must be a selected property, not simply a residue of properties that have not had time to change. Given the robust pace of evolution, conservation apparently does not impede diversification. The observation most demanding of further explanation is the stasis of core processes in the face of rapid change and divergence of anatomies and physiologies.

Phase 1: Novel Chemical Reactions

At some time more than three billion years ago, the last ancestor of all extant life arose, probably a bacterium-like organism. We now have microfossils of bacteria-like forms, 5–10 micrometers in length, in rocks three billion years old. Perhaps these are near relatives of the ancestor. What properties did that ancestor possess? It must have had all of the common characteristics shared today by modern organisms—say, humans, fungi, plants, and bacteria—since many of the chemical structures and chemical transformations are universally shared.[3]

We do not know whether life originated with RNA- or DNA-based heredity, or whether in fact heredity preceded or followed the evolution of proteins. Because all recent life forms contain DNA as the stable repository of the sequence information of proteins and use RNA as an intermediate interpreter of the DNA sequence, we can assert that around three billion years ago bacteria-like organisms were present that had DNA, RNA, a genetic code for 20 amino acids, and ribosomes as factories for making proteins under the direction of RNA. The basic processes of DNA replication, transcription into an RNA copy, and translation into protein had been established. This organism must also have been a self-replicating cell enclosed by an impermeable membrane of two layers (a bilayer) of lipids. It must have contained several hundred kinds of enzymes for synthesizing the major components of the cell, including the 20 amino acids, the cell membrane lipids, and the DNA bases. An energy metabolism based on the breakdown of sugars must have been established at that time. The synthesis of cofactors, which later became vitamins, would have been established as well. The organism of course would have had other attributes not commonly shared by its descendants.

The foodstuffs and energy sources for these early organisms may have been unusual—similar to those used by several groups of modern bacteria living in extreme conditions near boiling ocean vents. Carbon would have been available as carbon dioxide and sulfur as hydrogen sulfide from such geological sources, rather than as ready-made constituents of living matter or as carbon dioxide from the atmosphere. Oxygen was probably absent. Even if some of the sources of energy were different from those of modern bacteria, they would have already been very complex. Most of the biosynthetic pathways for making the 60 or so building blocks of cells would have been identical to those that now exist in all life forms.

These universal processes are the subject matter of courses in modern biochemistry and molecular biology. The chemistry of the processes was evolved at least three billion years ago; the components and their activities have been retained unchanged to this day, transmitted to all offspring of this ancestor. It is an amazing level of conservation. After these millions of millennia of evolution, many metabolic enzymes in the bacterium *E. coli* are still more than 50 percent identical in their amino acid sequence to the corresponding human enzymes. For example, of 548 metabolic enzymes sampled from *E. coli*, half are present in all living life forms, whereas only 13 percent are specific to bacteria alone.

The similarity is not just structural but functional. We can use modern recombinant DNA methods to exchange genes between very distant organisms, even bacteria and mice, after which they often still

	GEOLOGICAL EVENTS	BIOLOGICAL EVENTS
3 billion years ago	• formation of Earth • reductive atmosphere	• prokaryote fossils
		• bacterial radiations • cyanobacteria (O_2 producing)
2 billion years ago	• glaciation • accretion of continents • atmosphere slightly oxidative	
		• introns? • chromatin? • major protist radiations
1 billion years ago		• diversification of plants, fungi, and animals
	• increase in O_2 • breakup of supercontinent • glaciation	• segmentation • *hox* clusters
	• mass extinction • glaciation • mass extinction • warm earth • mass extinction • glaciation	• earliest jawed fish • land invaded by plants and arthropods • radiation of birds • radiation of mammals • earliest hominids
Present		

Figure 5 Timeline of geology, organisms, cellular innovation, and conservation. The evolution of organisms, exemplified here with animals, entails not only the diversification of anatomy and physiology but also the innovations and conservation of molecular components and activities, the so-called core processes.

CONSERVATION OF CORE CELLULAR PROCESSES

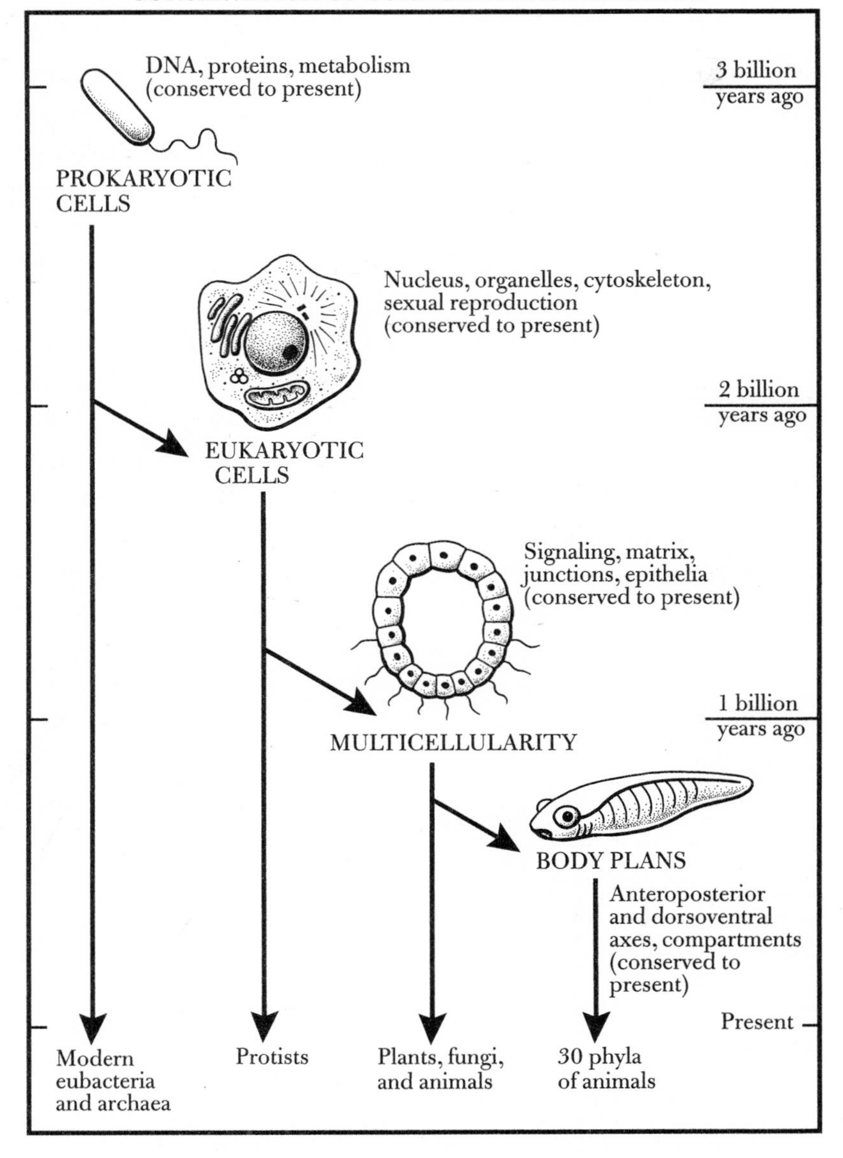

provide function. We will subsequently refer to these reaction series as conserved biochemical and molecular biological processes, the first set of the "conserved core processes" of living organisms. Chemistry of a biosynthetic and energy-yielding sort, and information retrieval from the genome, were achieved in this first phase of major evolutionary innovation in life. Once these processes were established, three billion years of stasis followed in these core mechanisms, right up to the present. Since all complex life followed, this biochemical stasis did not prevent the generation of novel phenotypes.

Evidence is completely lacking about what preceded this early cellular ancestor. Simpler organisms, such as viruses, are not free living; they are parasites on more complex forms and hence give no information on how the original bacteria-like organisms arose. In the hope of finding more primitive organisms that have retained ancestral characters, some biologists are actively exploring life forms that inhabit extreme environments on the Earth (hot springs, hydrothermal vents). To date they have found no clue to the earlier steps of evolution.

Another potential source of information is extraterrestrial. Francis Crick, codiscoverer of the structure of DNA, proposed that life migrated here as spores or bacteria that survived collision with the earth's atmosphere and surface.[4] There are plans to explore planets such as Mars and satellites such as Europa, a moon of Jupiter, or Titan, a moon of Saturn; these are sites possibly conducive to life as we know it on earth. Everything about evolution before the bacteria-like life forms is sheer conjecture, so we start this narrative with the bacteria-like ancestor and its complex collection of biochemical and molecular biological core processes.

Phase 2: Cell Organization and Regulation

The universal ancestor, it is thought, split into two major lines of bacteria-like organisms three billion years ago. One line led to the modern eubacteria, the other to the modern archaebacteria. The eubacteria form a large group of what we traditionally call bacteria, single celled and microscopic. It includes the pathogens that cause tuberculosis, syphilis, plague, and anthrax as well as many nonpathogens. They live in all environments of the earth and are extremely versatile in their capacity to make and destroy chemical compounds. The archebacteria are particularly interesting, because eukaryotic organisms and ultimately human beings arose from them. Today the archebacteria mostly inhabit extreme environments. One large group produces methane, another lives in salty environments, and a third lives in hot springs, or in nearby sulfur deposits, or near hydrothermal vents in the ocean.

For many years, before DNA sequencing was possible, these two major groups of bacteria were not distinguishable by appearance; we did not appreciate how different they are. The commonalities of these modern bacterial groups tell us about the cell organization of the ancestor of all life. It must have been single celled and microscopic, the smallest living cell measuring 1–5 micrometers across, with its DNA contained within the cytoplasm, not enclosed in a nucleus. This cell organization is shared by all modern bacteria. All of these forms, as illustrated in Figure 6, lack a defined cell nucleus and are called prokaryotes.[5]

Among the eubacteria are the cyanobacteria (once misleadingly called blue-green algae), which invented oxygen-generating photosynthesis by which the entire oxygen atmosphere of the earth was slowly produced from water. Thanks to their work, oxygen probably increased to a few percent in the atmosphere by two billion years ago (presently it is 21 percent).[6] At roughly that time a second phase of innovation occurred: the archaebacterial line split into two lines, one leading to the modern archaebacteria and the other to eukaryotic organisms (organisms which contain a nucleus). These now include diverse single-celled protists (what used to be called protozoa, for example, an amoeba or paramecium) and the great multicellular kingdoms of plants, animals, and fungi.

Eukaryotic cells differ greatly from prokaryotic cells, and enormous innovations attended the evolution of the first single-celled eukaryotes one and a half to two billion years ago. These innovations mostly concern more complex cellular organization, as shown in Figure 6. Today eukaryotic cells from sources as different as yeast (a fungus),

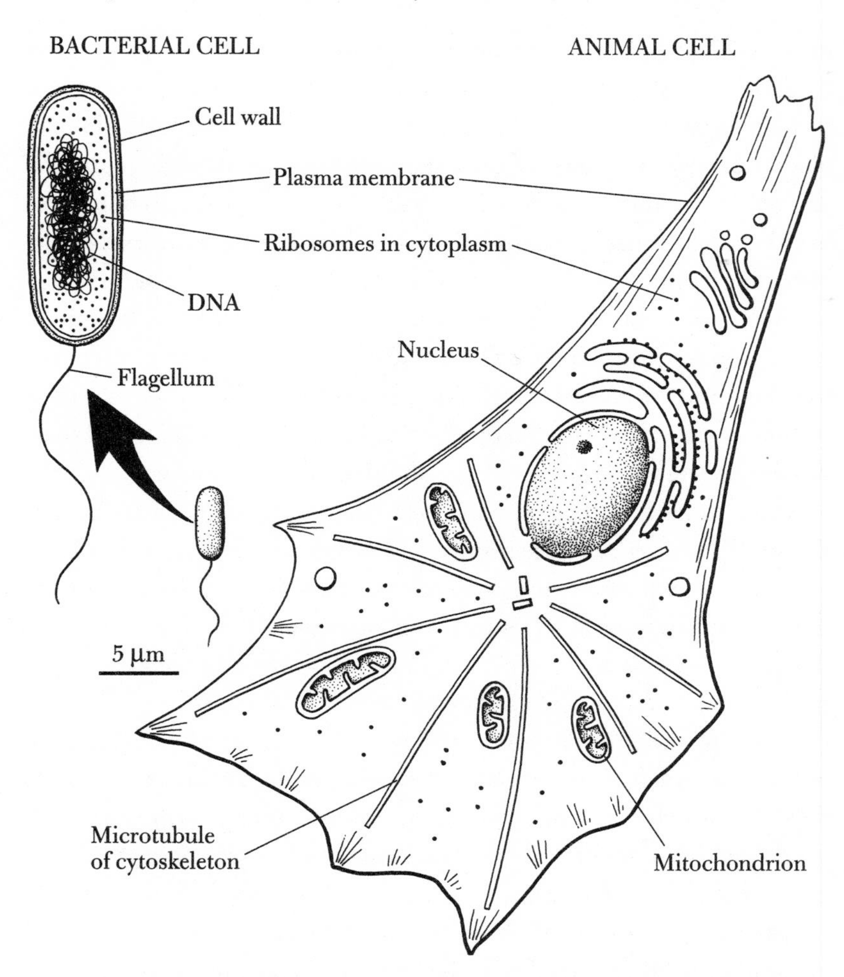

Figure 6 Prokaryotic and eukaryotic cells. *Left*, prokaryotic cells, which first appeared perhaps three billion years ago. Modern bacteria have prokaryotic organization. Notice the lack of internal compartments. *Right*, eukaryotic cells first appeared perhaps two billion years ago. Notice the large size, internal compartments including the nucleus, and the cytoskeleton.

protists, plants, and animals are very similar in their organization, from which we can conclude that their common ancestor evolved an extensive set of traits now common to all descendants. Generating the first eukaryotic cell was a major and enduring accomplishment. Because of the similarity between human genes and those of more primitive organisms, these simpler organisms can be used as models for the discovery of drugs, which can be quickly applied to human beings.

A suite of features was involved in the "invention" of eukaryotic cells. The most striking trait is their size and complexity. They are one hundred to one thousand times larger in volume than bacterial cells and have numerous internal membranes that wall off small compartments or organelles ("little organs"), which are specialized for different functions. The DNA is located in one such organelle, the nucleus, in which all of the RNA is copied from the DNA. Such specialization of the cell's volume was limited in the small prokaryotic cell line. Eukaryotic cells contain larger quantities of DNA than prokaryotes. For example, the gut bacterium *E. coli* has 4,300 genes, whereas yeast has 6,300, and humans have about 22,500. The amount of DNA not coding for RNA, sometimes called junk DNA (a dangerous term for something one does not understand), is also much greater in eukaryotes. Thus, the ratio of genome size to the number of genes is one hundred fold greater in complex animals than it is in bacteria.

Eukaryotic cells developed the capacity to engulf large food particles, whereas bacteria, each surrounded by a rigid wall to maintain shape, secrete digestive enzymes in their environment. In the wall-less eukaryotes, the cell shape is maintained by an extensive internal cytoskeleton, which is dynamic and can take on various configurations, and by pumps for small ions like sodium which keep the cells from bursting owing to osmotic pressure. As they engulf food particles, eukaryotic cells enclose them in a membrane vesicle and transport the vesicles through the cytoplasm. These vesicular compartments are directed to special uses within the cell by fusing them to other vesicular compartments containing digestive enzymes. Various organisms have changed the number, nature, size, and function of the compartments, but the basic rules for establishing them and for moving materials between them must have evolved in the earliest eukaryotic lineages nearly two billion years ago.

A crucial conserved feature of all eukaryotic cells is sexual reproduction. Bacteria rarely exchange genetic information, whereas most eukaryotes do so frequently by way of sexual reproduction. In some lines it is a required process of every life cycle. Sexual reproduction is not indispensable; various eukaryotic organisms can reproduce either asexually by dividing in two, or sexually whereby two cells, having reduced their chromosome content to half during meiosis, fuse together to restore the full chromosome content, and then divide. In some multicellular organisms the two fusing cells are very different, as in egg and sperm, but in most single-celled eukaryotes they are similar. Sexual reproduction has the same basic design in all eukaryotes, and it must have evolved in the days of single-celled ancestors.

Extensive innovation showed up in the complexity and organization of the eukaryotic ancestor: much more compartmentation of its components and reactions, much more spatial organization, much more regulation of when and where events occur in the cell, and more temporal specialization of the different phases of the cell cycle. These features would later be widely exploited in complex multicellular organisms.

The eukaryotic cell organization also spawned an incredible diversity of protists, all single celled, in the last billion and a half years—even without considering the path that led to multicellularity. Yet on a metabolic level the early eukaryotic cell was rather simple, probably obtaining most materials ready-made from the bacteria it ate. The innovations have been conserved and carried forward to all modern eukaryotes without change. The commonalities of eukaryotic biology constitute the content of university courses on Mendelian genetics and on modern cell biology. They are conserved genetic and cell biological processes, and they form the second group of conserved core processes of living eukaryotic organisms, including humans.

Phase 3: Multicellularity

By 1.2 billion years ago, the single-celled eukaryotes had diversified to a lush variety of forms, one of which underwent changes conducive to the acquisition of multicellularity. That unknown ancestor has evolved into all multicellular eukaryotes—namely, the kingdoms of plants, animals, and fungi. In one line, a descendant leading to the plants engulfed a cyanobacterium and achieved photosynthesis in one step. In another line of descendants leading to the fungi, an ancestral cell with a strong wall gained great metabolic versatility, rivaling that of bacteria, and lived on ready-made foodstuffs from other organisms, living or dead. (The degradative capacity of fungi is legendary; for example, they can live off synthetic chemicals like PCBs.) In a third line, the ancestral cell omitted a rigid wall and remained a complex feeder, needing a diet of whole cells to supply it with a variety of prefabricated amino acids and vitamins. It evolved into the animals (metazoa). All the multicellular lineages are diverse; fungi alone comprise over one hundred thousand known species.

In the early period of multicellularity in the animal lineage, this ancestor gained many new properties related to the much more social lifestyle of the cells of which it is constituted. It evolved proteins that allowed cells to stick to one another. A key event was the assembly of the epithelium, a closed sheet or sphere of cells, as described in Figure 7. Cells of the epithelium have complex junctions so tightly welded together that virtually nothing can pass between them. Thus the pumps and channels in the cells themselves control the salt composition of the milieu inside the sphere, making it hospitable compared to the outside. Metazoans also evolved a secreted matrix (called the extracellular matrix) upon which the epithelium sits, as well as specialized proteins that allow cells to attach to the matrix. The matrix gives the epithelium greater strength. A component of this matrix is the protein collagen, which has the same structure from sponges to humans, conserved from the earliest animals.

A controlled fluid environment inside the multicellular epithelial organism was a novelty that promoted communication between animal

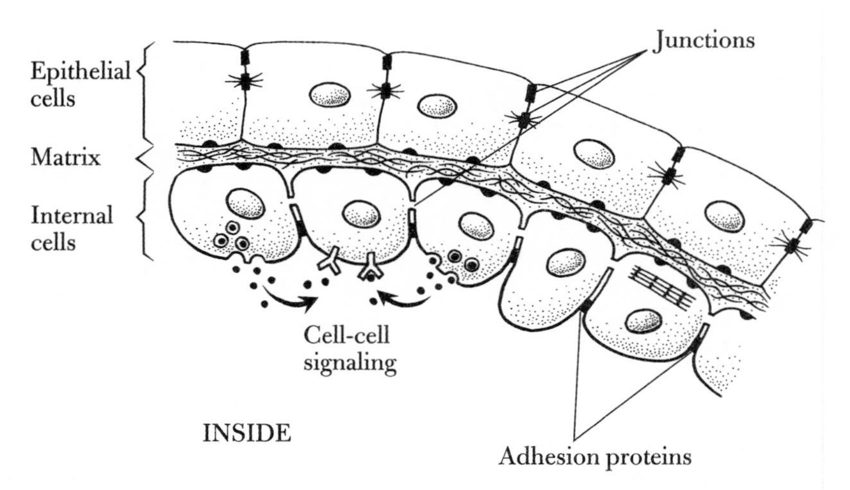

Figure 7 Innovations in multicellularity. The first eukaryotic multicellular organisms appeared perhaps one billion years ago. They possessed innovations of cell communication, cell contact, and cell differentiation.

cells via secreted and received signals. Communication, of course, also occurs in single-celled eukaryotes and prokaryotes, but not to the same extent as in metazoans. Other eukaryotes such as multicellular algae, slime molds, and mushrooms, lacking intercellular junctions, nonetheless achieve sufficient intercellular signaling for elaborate multicellular structures, such as the fruiting bodies of mushrooms. Plants, for example, have sizable channels that allow large signaling molecules, even RNA molecules, to pass between cells. The controlled internal milieu of animals, though, must have provided the context for the elaboration of a greatly expanded set of signals and receptors, and indeed animals have evolved many kinds of cell-cell signaling.

These new multicellular animals retained the sexual mode of reproduction, which required in each life cycle a return to the single-celled state, the fertilized egg. Development from the egg restored the multicellular state. Elaborate development is a phenomenon of multicellular forms. Cells gained numerous differentiated functions, such as the contractility of the muscle cell. Yet contractility was not an innovation that emerged entirely from nothing. All the components and processes had already been used in ancestral eukaryotic single cells such as contractile proteins, the structural scaffold proteins on which the contractile proteins operate, and the calcium-based triggering system.

These conserved components were all brought forward in the newly evolving animals, but were produced at high levels and reorganized for greater contractile efficacy in this one kind of specialized cell. Similarly, the degradative activity of differentiated digestive cells was not an innovation but an exaggeration and reorganization of the digestive capacity of ancestral eukaryotic single cells. The highly specialized nerve cell coupled aspects of single-cell function into a novel transmission system.

The evolution of differentiated cells was a regulatory accomplishment involving new placements and increased amounts of old components. Once evolved, many of these cell types were conserved in metazoan evolution, from jellyfish to humans.

With the evolution of developmental processes came a general and complex life cycle of animals. At some early time in metazoan evolution, the germ line of cells separated from the somatic lines—the former capable of inheritance, the latter of development and differentiation but not inheritance. All of this complexity was achieved in a period from a billion years ago to approximately 600 million years ago, still before the Cambrian period. The newly originated processes and the functioning components constituting them have been conserved in all living animals, including sponges, insects, snails, and mammals. They must have been present in the early multicellular ancestors before the phyla diverged. They are the conserved multicellular and developmental processes, another subset of the conserved core processes of animals, these arising a billion years after the genetic and cell biological subset.

Phase 4: The Origin of Body Plans

By 600 million years ago, fairly complex animals were probably present: branching sponges, radial animals such as jellyfish, and the first small bilateral animals (like us, with mirror-image left and right sides), perhaps rather worm-like in form, which left traces of their burrows in the muddy ocean floor, thereafter fossilized. Many biologists think that the worm-like ancestor must have descended with modifications from a radial ancestor, which resembled in many ways a modern coelenterate (such as jellyfish) or ctenophore (comb jelly). This worm-like animal may have been the ancestor of all modern bilateral animals.[7]

Rather suddenly, diverse macroscopic anatomy appeared on the Cambrian scene of 543 million years ago. By the Midcambrian, representative animals of all but one of the 30 modern phyla were present, according to fossil records (especially from the Burgess Shale in Canada and the Chengjiang formation in China). For example, mollusks (clams and snails), arthropods (insects and crustaceans), annelids (earthworms and leeches), echinoderms (sea urchins), and chordates were present. Even fish-like vertebrates have been recognized. Many macroscopic forms were well fossilized; for instance, animals with shells and tough cuticles, many more than one quarter inch (1 cm) in length, which is enormous compared to most unicellular protists and bacteria. All of these macroscopic complex forms may have descended from the worm-like ancestor.

The abruptness of the emergence of so many complex anatomies may be an artifact of the special features of fossilization at that time, or of some special environmental condition that favored large and more complex animals, or it may be the result of some breakthrough in regulatory control on the cellular level. Once again, a new suite of cellular and multicellular functions emerged rather quickly and was conserved to the present.[8]

We will try to infer the character of the ancestor of all bilaterally symmetric animals, not from fossils but from the shared and conserved anatomical and molecular properties of modern bilateral animals descended from it. It would indeed be informative to find Precambrian fossil remains of the ancestor, but none have been found. Two great Precambrian glacial periods occurred about 620 and 580 million years ago, and these may have interfered with fossil deposition. Nonetheless, we can surmise that the bilateral ancestor had a through-gut (two openings) rather than a blind gut (one opening), as do coelenterates like hydra and jellyfish. It possessed a head at the anterior (mouth) end, which had a concentration of sense organs and nerves. Its body was concentrically organized into three layers rather than the two possessed by ctenophores and coelenterates. The new middle layer, which is absent from modern radial animals like coelenterates, formed muscle. Thus, the bilateral ancestor had already made great anatomical strides from its radial ancestor.

Whereas these shared anatomical features were recognized a hundred years ago, comparisons of the past fifteen years have revealed unsuspected shared aspects of development. The embryos of animals as anatomically divergent as fruit flies and mice express similar genes in similar places; a grid or map locates the development of specialized parts at different places in the body. This map reveals a hidden anatomy where seemingly homogeneous tissues are subdivided into discrete territories. A rather similar map must have been present in the worm-like hypothetical bilateral ancestor, and then conserved to the present except for small changes, because all bilateral animals share major features of this map.

By deduction from the maps of living descendants, the body of the bilateral ancestor was probably subdivided into five to ten large domains in the head-to-tail dimension and a few more in the back-to-belly direction, producing a checkerboard of unique regions. The ancestor probably had a heart-like pumping organ (common to many but not all bilateral animals), anterior light-receptive cells, and a complex nervous system perhaps centralized into a structure at the back or in the belly, or more likely diffuse around the whole body. To judge from the width of the trace fossil burrows, this ancestor was probably less than one tenth inch wide, perhaps only 1 mm.

By the Midcambrian, about 30 different phyla of bilateral animals had evolved from this ancestor. Each phylum is distinguished by a body plan—a unique global body organization—so this number of body plans had evolved by the Midcambrian from the worm-like ancestor. The body plans of three major phyla—chordates, arthropods, and annelids—are shown in Figure 8. These plans have been conserved and carried down to the present. No new body plan has evolved since

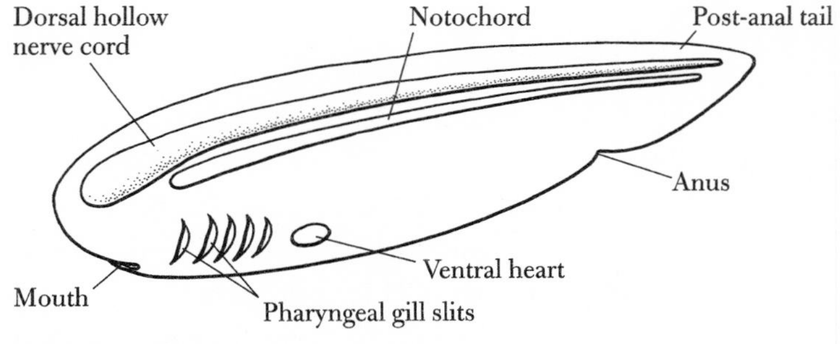

CHORDATE BODY PLAN

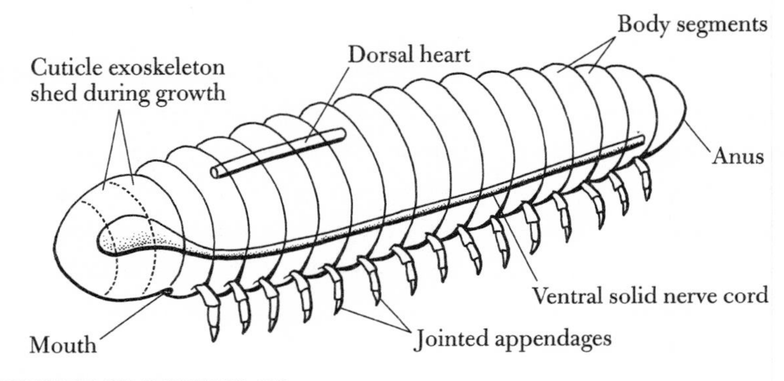

ARTHROPOD BODY PLAN

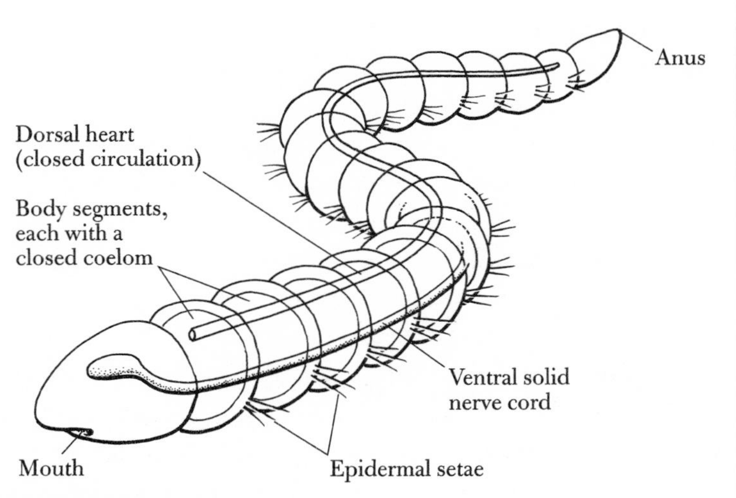

ANNELID BODY PLAN

Figure 8 Invention of body plans. Complex bilateral animals evolved in the Precambrian period more than 545 million years ago. Representatives of most modern phyla were present in the Cambrian, recognizable in fossils by their unique body plans.

then, except perhaps that of the phylum Bryozoa in the Ordovician (450 million years ago). The period just before and during the Cambrian was the phase of innovation of body plans, the fourth phase of our series. The worm-like ancestor must itself have had a rather complex body plan organized according to the traits mentioned above, and the 30 new plans were modifications thereof.

Innovations of the Body Plans

We will follow the lines of descent with modification from the bilateral ancestor to three phyla. In the line to the arthropods (insects, crustacea), which displayed enormous variety already in the fossils of the Cambrian, the body plan of the bilateral ancestor was modified to add body segments, appendages to each segment, and a tough outer layer shed regularly after intervals of growth. The nerve cord may have condensed along the belly. Various gene expression domains of the map of the worm-like ancestor were kept; they may have been enriched and modified, but the domain map did not change much. The arthropod body plan was thus devised. The animals were highly motile by virtue of their jointed appendages and jointed body.

In the line to annelids (worms), segmentation would have been added, probably independently from that of arthropods. A lined body cavity was formed in every segment. The nerve cord condensed along the belly. As in the arthropods, the various gene expression domains were kept from the map of the ancestor with some enrichment and modification (different from that of arthropods), but were not changed much. Thus, the annelid body plan was devised. These animals might have moved by sinuous swimming and pulsatory burrowing of the entire body, much like present-day earthworms and marine worms.

In the line to chordates and then to vertebrates, segmentation would have been added, independently invented yet again, in the muscle blocks that parallel our spine. In some groups, like snakes, a large number of these blocks would have formed. Gill slits would have been added even before chordates arose, then they were conserved and carried into the chordates. The nerve cord condensed along the back. A rigid rod developed internally from the roof of the gut against which swimming muscles worked. The various gene expression domains of the ancestor's map were kept, with some enrichment and modification (different from that of arthropods and annelids). A tail was extended beyond the anus. Through these intermediates, the chordate body plan arose. Movement was by sinuous flexion of the body, with the tail for added motility.

A similar story can be formulated for the anatomical modifications of the ancestor toward each of the 30 bilateral body plans of each of the 30 bilateral phyla, completed more than a half-billion years ago. Each phylum has a different embryology with respect to the generation of these different body plans. Then, as the body plans were established, each became conserved and inherited with rather little modification by all subsequent members of the phylum. Conservation was the rule once again, now seen at the anatomical level of the whole body.

Although the body plan is an anatomical structure, it plays a central role in development, and it too should be called a conserved core process. It joins conserved processes such as metabolism and other biochemical mechanisms, eukaryotic cellular processes, and the multicellular processes of development to make up the repertoire of conserved processes of bilateral animals.

According to Eldredge and Gould, individual species in the fossil record often go through a period of anatomical stasis as part of punctuated equilibrium, but such stasis would be difficult to recognize in all the nonanatomical aspects of the phenotype that did not fossilize, such as physiology, behavior, and development. Yet it is undeniable that the conserved core processes of animals, based on their broad commonality, have not changed for long periods, whereas other features have evolved rapidly. The progression or "moving front" of biochemical, then cell biological, then multicellular, and then phylum-specific embryological mechanisms has repeatedly involved one explosion of novelty after the other. Each explosion was followed by enduring stasis, while the individual animal species evolved rapidly on other fronts—those of anatomical and physiological additions to the body plan.[9]

Phase 5: Fins and Limbs in Chordates, Appendages in Insects

The period since the Cambrian seems to have been one of rampant anatomical diversification at different sites on the body plan, while the plan itself remained unchanged. Vertebrates, a subgroup of the chordates, were first recognized in the Cambrian. By the Ordovician these, our ancestors, were armored jawless fish, a few inches in length, moving slowly by tail propulsion. They probably fed by drawing a current of water into the mouth cavity and forcing it out the gill slits. From there, particles were collected on a sticky track on the floor of the mouth, which moved to the gut. The later-evolved jawed fish became efficient predators.

Adaptation for predation represents a substantial modification, in which the jaws supported biting and active hunting, and the body was capable of undulatory movement. The armor was shed. These early fish may have been stabilized in their swimming by two flaps extending laterally for the length of the body, like two long fins. Probably the first paired fins, anterior and posterior pairs, were derived from these continuous fins by the time of the Silurian period (443–417 million years ago). Paired fins were of great importance in balance (pitching and yawing) and control of direction for these faster-moving fish, which radiated from shallow-water into oceangoing forms.[10]

One line of fish, the sarcopterygians, which was present by 400 million years ago, had fleshy stumpy lobe-fins. To judge from fossils and from still-living relatives (the coelacanth and lungfish), the fins probably contained a linear series of bones, concentrically surrounded by muscle, nerve, dermis, and a surface of tough epidermis. At the tip of the fin were thin rays of bone supporting the swim fan. Various kinds of these lobe-fin fish probably lived in shallow seas and lagoons, close to shore. They may have waddled around in the shallows on their stubby fins, preying on fish and rich refuse from shore, for plants and arthropods had already been on land for millions of years. The lobe-fin fish 40 to 80 inches (1 to 2 m) long were the largest predators in this milieu.

In the late Devonian period, 380–360 million years ago, within a

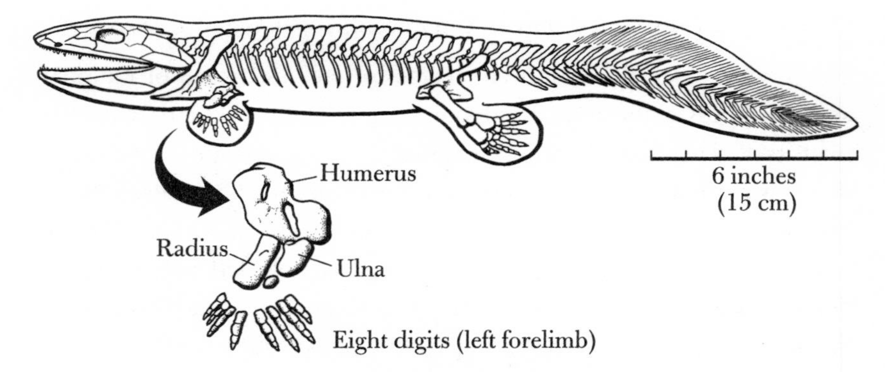

Figure 9 The origin of terrestrial limbs. The first land vertebrates arose over 360 million years ago. Their limbs were modified fins with novel additions, the wrist and hand parts. *Acanthostega,* an amphibian of 365 million years ago, is depicted, showing its left forelimb with eight digits.

line of sarcopterygians arose a modification at the tip of the lobe-fin. The fan disappeared and the autopod, which has become the wrist and hand (or ankle and foot), formed in its place. The digits are quite novel and do not seem derived from the fin ray bones. For short periods the descendants came onto land, where food was probably abundant. These animals were increasingly "amphibians," living both on land and in water. The number of digits at first was variable. Some early amphibians had seven, eight, or nine, as illustrated in Figure 9. Others had five. Eventually the number settled at five in the common ancestor of all modern land vertebrates, such as ourselves.

Over subsequent eons, the various parts of the limb differentiated further, as did the fore and hind limbs from each other. In some lines, they broadened at the end as flippers; or curved and fused as diggers for burrowing; or fused and lengthened as long legs with hooves, or webs for swimming, or dexterous individual digits in the apes, or enlarged surfaces as wings. Wings arose at least three times: in the pterodactyl reptiles, the birds, and the bats. Different digits were used in each to support the broad wing web.

Even the most unusual limbs derived from the ancestral fish fin still have serial bones in the center, and muscle, nerve, dermis, and epidermis around. The modifications involve mostly rearrangement, repetition, change of size, and branching of old parts. The modern fish fin and the tetrapod limb share many aspects of development and organization. This developmental module arose with the jawed fish about 400 million years ago and has been conserved ever since. It joins the other accumulated processes in the repertoire of conserved core processes of land vertebrates. When the wrist and hand evolved into the five-digit hand, the period of innovation was followed by conservation of that innovation too, through to the present in land vertebrates. Today every terrestrial vertebrate limb is built on a variant of the ancestral five-digit limb, which has specialized further.

Arthropods have also undergone an extensive evolution of their appendages by way of modification of an ancestral protrusion. The body of an arthropod is divided into a series of segments in the posterior head, thorax, and abdomen, each allowing a small amount of flexure at the joint. The anterior head may have been segmented as well, although in modern members the flexible joints have disappeared in favor of a fused unit. Ancestral arthropods of the insect-crustacean line, while still oceangoing, had a leg-like appendage projecting from each segment, so the elongated animal was rather like a centipede. The legs had two parts: a feathery respiratory fan on the upper part and a jointed leg on the lower part. Such a two-part structure is still found on many crustacean segments. All segments in this ancestral arthropod had a rather similar appearance, as shown in Figure 10.

The early insects, which arrived on land in the Silurian and Carboniferous, about 400 million years ago and long before the vertebrates, are thought to have lived at first as flightless consumers of decayed organic matter. As time went on, the appendages of the various body segments gained exquisite individualization of function by way of modifications of their basic tubular and articulated structure.

In the insect line, the appendages of the head became truncated and modified into various mouthparts for clasping and cutting; the antenna was modified further with various sensory structures for special uses. Legs on the thorax became longer and differentiated, and the abdominal legs were suppressed entirely in the adult. The true

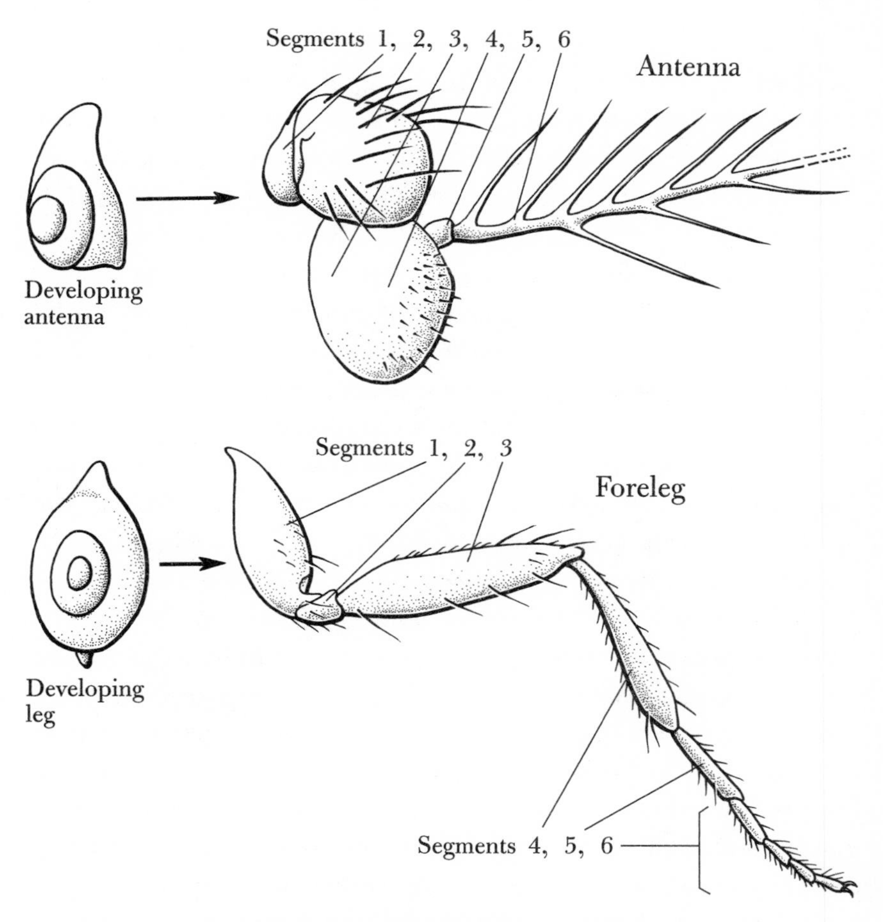

Figure 10 The common plan of insect appendages. The antenna and leg are tubes having the same number of jointed sections. They develop from similar nests of cells, which telescope outward to form the appendage.

insect adult has just six legs, all on the thorax. The most posterior appendages were modified into the clasping organs of mating. Interestingly, all appendages including the antenna share numerous aspects of early development. There seems to be a conserved developmental module common to all appendages of all insects. Then, in the individual appendages in the lines of insects, the process was modified and supplemented, to give great variety.

Wings evolved in some Devonian insects, it is thought, as modifications of the feathery respiratory fan located above the ancestral tubular leg (but originally part of the leg). Ancestral insects of this line had wings on many segments of the posterior head, thorax, and abdomen. Later, wings were suppressed in all segments except two of the thorax. The enormous dragonfly-like insects of the Devonian had four simple wings of two-foot span. Eventually, wings became more differentiated in size and shape. In virtually all two-winged insects, such as fruit flies, the posterior pair has been suppressed to stubs, which serve as balancing organs, leaving only two flying wings.[11]

Looking at these changes, one has the distinct impression that the modifications involve rearrangement, local shortening or lengthening, or truncation, but are not anything basically new. On the other hand, each morphological change has had vast implications; for example, in the feeding range of insects.

One of the major events of the Carboniferous, it is thought, was the evolution of wood-boring and leaf-chewing (phytophagous) beetles showing modifications of appendage-derived mouthparts. The newly evolved large, woody, tough-leafed trees of the Carboniferous became their targets. The elaborate modifications of anterior legs and antennae in conjunction with the emergence of flowers in the Cretaceous opened new vistas of pollination, plant-insect coevolution, and nectar feeding. But behind each modified appendage remain the common conserved processes of appendage development that were invented in the Cambrian.

The Duality of Conservation and Diversification

If we follow the path from the bacterium-like ancestor toward humans, we find repeated episodes of great innovation. New genes and proteins arose in each episode. Afterward, the components and processes settled into prolonged conservation. The existence of "deep conservation" is a surprise. To some biologists it is a contradiction of their expectations about the organism's capacity to generate random phenotypic variation from random mutation. To some, it borders on paradox when held against the rampant diversification of anatomy and physiology in the evolutionary history of animals.

Widespread conservation must reflect some limit on the organism's freedom to generate viable variation in all directions under the impact of mutation. If phenotypic variation is really smoothly continuous in all directions, then all components should vary and nothing should be conserved for very long. When similarities are found, they may mean only that the two descendants have had a common ancestor so recently that they have not yet had time to lose all the originally shared traits. That is clearly not the case for the conserved core processes. These are brought forward to the present in the lineage as functional modules from each epoch of cellular innovation.

The conserved components and processes are certainly not just random remainders on the way to alteration, but are integrated functioning pathways and circuits, the core processes of the organism by which the phenotype is generated from the genotype. They are the essentials of synthetic and energy-generating metabolism, of the development by which the anatomy is generated, and of the organism's physiology.

What about the core processes is conserved? Based on extensive comparisons of DNA sequences over the past 20 years, a deep conservation of coding sequences, those encoding the amino acid sequence of proteins, is incontrovertible. For example, many enzymes of metabolism that are components of some of the core processes shared by bacteria and humans are conserved, though these are separated by three billion years. Conservation cannot get much deeper. Their protein functions and their encoding DNA sequences are conserved.

What is conservation attributed to, and what is it not attributed to? If human and bacterial metabolic enzymes have nearly the same amino acid sequence, it is not because the coding DNA has been especially protected from random mutation. Sequences in the DNA that can change without altering an encoded amino acid are extensively changed, reflecting the numerous random mutational hits. If the mutational change has no consequence, then it is free to remain in the genome of the offspring. It is silent or neutral. At other positions,

those changes that do lead to an amino acid change must have been eliminated as lethal or reproductively disfavored over time.

Hence, except for silent changes of the DNA sequence, the coding sequence of these ancient proteins remains unchanged. While all genotypic variation is possible in the sense that every base position of DNA can be changed, only some phenotypic variants are viable. Those mutations in the DNA that substitute different amino acids into areas of the protein with critical functionality will generally lead to inactive proteins. The result will be lethality in the organism and their quick elimination from the population.

Darwin's supposition that change is pervasive has to be replaced with the view that in the history of life some things change and others do not, and that change occurs in spurts and then becomes fixed and subsumed in all descendant organisms. The eons of evolution have seen a moving front of episodic, innovative additions of core processes. The fixation occurs in important core processes that provide metabolism, information retrieval, signaling, and developmental mechanisms. Superimposed on top is ongoing anatomical and physiological innovation.

Why do improvements in the core processes, which must have been invented in brief episodes, stop accruing during the long periods of stasis? Perhaps most surprising are the body plans (phyla) that emerged anew in the Cambrian and were conserved, perhaps only one arising thereafter. Why did inventions of body plans cease?

Body plans at first were probably unadorned. But it is likely that during the Cambrian, when the 30 or so existing body plans were in the process of being fitted with armor, biting parts, and appendages, the *new* unadorned body plans, even if they were of improved design, might be particularly vulnerable. These new "phyla" were simply lunch for the highly protected and aggressive established phyla. By the Cambrian, the locus of battle had shifted from who could make the best body plan to who could make the best jaw and appendage on an adequate body plan. The days of body-plan wars had ended; a different technology was at stake. We can imagine that similar arguments of preemption would hold for stasis in the other core processes.

After the fixation of core processes, evolution seems to have proceeded at a steady if not increasing pace. The conserved processes are called core processes because they are deeply involved in generating the phenotype. Their constraint on changes of amino acids is due to the likely impairment of their function. Mere inability to change does not explain their long-term persistence, because other processes could in principle arise, surpass, and replace them. Why do they persist for such long evolutionary times? Are they continually under selection for the properties they have? Or have they reached some sort of optimality, where any change would be for the worse? Is it because they are deeply embedded in so many other processes (although this bypasses the question of why they are embedded)? If they are under continuous selection, what are they being selected for? These far-reaching questions raised by the evolutionary history of adaptive cell behaviors are immediately relevant to the facilitation of variation and the pace of evolutionary change.

To go still deeper, why did the history of life on earth progress in this way instead of in the way Darwin surmised, where every aspect of the organism would be subject to change? Is inhomogeneous change somehow more effective than homogeneous change, so that the existence of conserved core processes was an inevitable outcome of evolution? Core processes that transformed the organism seem to have been invented only a few times, in episodes, on the way to multicellular animals. Can we predict theoretically that biological organization based on these features, though hard to achieve, is more effective in generating phenotypic variation than pervasive and continuous change? We will return to those questions after we have examined the core processes themselves and their special adaptive properties.

THREE

Physiological Adaptability and Evolution

Evolution, as we have seen, is framed by two features: conservation on a cellular level and diversity on an anatomical and physiological level. How does diversification occur despite so much conservation? In this chapter we examine a few examples that reveal how conservation actually enables variation. The connection between the two is at the level of mechanisms; those that are exploited for evolution are the very ones that the organism uses day to day to vary its phenotype to meet new physiological demands. Such mechanisms can be easily modified in evolution to yield new phenotypes.

The potential relationship between physiological variation and evolutionary variation had been considered by some evolutionary biologists in the premolecular era, without wide acceptance. It is on the molecular level that the link between the two is seen most distinctly and where the evidence for facilitated variation is most persuasive.

Physiological Variation and Evolution

In the nineteenth century, scientists and philosophers struggled to settle the issue of whether the organism could pass its somatic adaptation, its so-called acquired characteristics, to the next generation. As attractive as that idea was to Lamarck and even to Darwin, it was decisively nullified both experimentally and mechanistically. Somatic adaptations include the physiological, behavioral, anatomical, and developmental changes that take place within a generation, are made in response to environmental changes, and are directed to the organism's immediate benefit. They are often reversed as the environmental challenge subsides. Evolutionary adaptations, on the other hand, are heritable changes of physiology, behavior, anatomy, or development to the organism's immediate benefit, transmitted over many generations and lasting even when the environmental challenge is gone. Although sometimes occurring under similar environmental conditions, somatic and evolutionary change seem very different from each other, and there is no known mechanistic path from one to the other.

Despite the high barrier erected by Weismann between somatic change and evolutionary change (not the least of which is that the former occurs in somatic cells and the latter in germ cells), a few biologists subsequently sought a new relationship between the two. They realized that what is selected is not simply a specific state of a biological system, but more commonly mechanisms that can produce a range of states in response to a range of conditions.

For example, two organisms may differ not only in their optimum temperature or optimum salt concentration but also in the range of temperatures or salt concentrations they can tolerate. The human body is selected not only for a certain level of muscle performance but also for an ability to change that performance under medium-term environmental stress, as shown in Figure 11. Organisms that live in constant environments generally have a narrow range of conditions for viability, whereas those that endure more varied environments have evolved mechanisms to tolerate wider ranges.

More and more studies in medicine point to sophisticated physiological mechanisms that could only have evolved under selection, whose sole function is to extend the tolerable range of conditions in which human beings can survive. For example, human beings sweat to cool off in the heat and shiver in the cold to generate heat by increased muscle contraction. They have several means to acclimatize to high altitude or physical exertion. Our bodies regulate and adjust

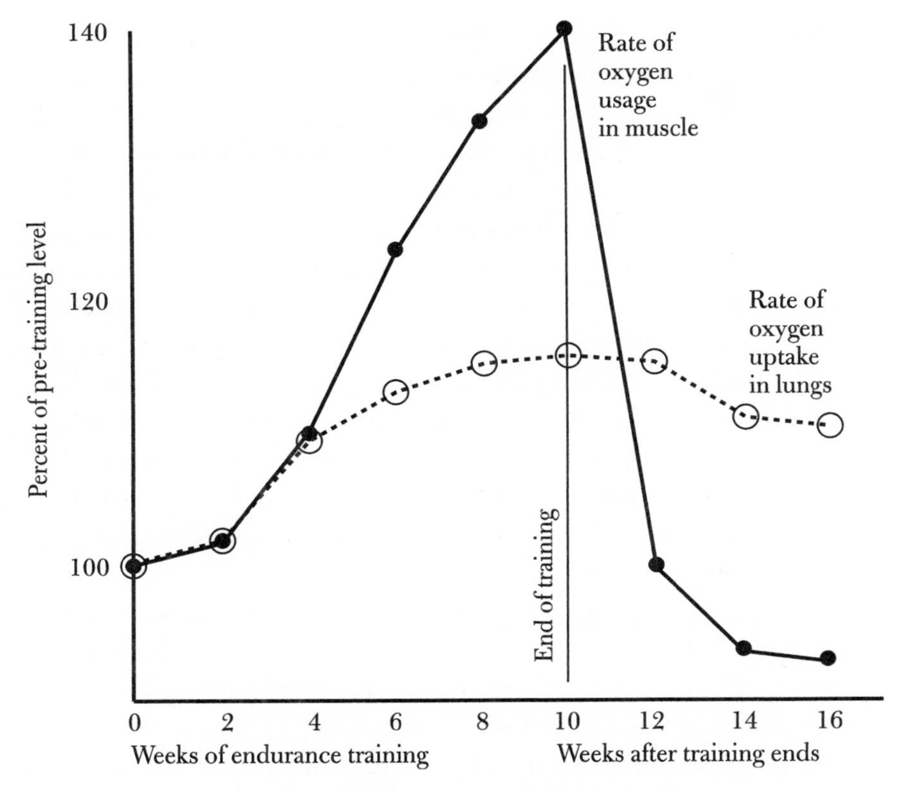

Figure 11 Human adaptation to fitness training. Thirteen people entered a ten-week endurance training of distance running and moderate weight lifting. Leg muscle was tested at intervals for its capacity for oxygen-dependent energy production (cytochrome oxidase). Note that the body adapts by raising its energy production by 40 percent, whereas oxygen uptake in the lungs increases only 15 percent. Deadaptation is rapid after training ends. (Redrawn from J. Henriksson, and J. S. Reitman, "Time course of changes in human skeletal muscle succinate dehydrogenase and cytochrome oxidase activities and maximal oxygen uptake with physical activity and inactivity," *Acta Physiologica Scandinavica* 99:91–97, 1977.)

blood sugar, blood pressure, caloric input, and ionic balance over a spectrum of environmental conditions.

Other biologists saw that somatic adaptation could feed back on evolutionary change by improving the organism's viability under stressful conditions. Somatic adaptability in effect creates a range of alternative phenotypes on which selection can act. Human beings display three interconverting types: the normally adapted animal, the sweating animal, and the shivering animal. The increased viability enabled by these mechanisms for temperature compensation can put the organism under new selective pressures. For example, increased ambient temperature would be expected to select for mechanisms for better adapting to heat. This extended range might also bring with it other selections caused by changes in the food supply, parasites, or predation. The extension would allow at least marginal adaptation under a wider range of environmental conditions.

Since we know that natural selection is better at improvement than creation, selection can occur efficiently on one of the extreme states, which many if not all members of the population could occupy. Under tolerable but stressful conditions, there would be strong selection for any genetic changes that would produce better adaptations, that is, a still more reproductively fit individual. This process is much more rapid than one in which the population of organisms is unable to occupy a new niche at all and simply waits for new mutations to arise one at a time. The extended range allows for extended selection and, as we shall see in this chapter, it offers improvements in selection without assuming that somatic adaptation to the environment can be directly converted to stable evolutionary changes.

Though the idea that evolution is based on physiological adaptability was first introduced more than a hundred years ago, its implications are being seriously revived from several perspectives. Basing evolutionary variation on somatic variation greatly facilitates evolutionary adaptation by allowing the organism to simply stabilize elements of existing phenotypes through new mutations. That said, somatic adaptability provides for only certain types of facilitated phenotypic variation.[1]

There is a very different sense in which somatic adaptability and evolutionary adaptability might be closely related, one that is only just discernible on a molecular level. Organisms have found simple and elegant solutions for altering their phenotypes in response to environmental or developmental stimuli. These same mechanisms might be particularly important targets for genetic modification in evolution. The hard work of creating alternative states was already done before these processes were modified genetically. The capacity to be regulated was inherent in establishing somatically adaptable mechanisms in the first place. The easier task is finding the modifications that would stabilize part of an already adaptable phenotype. The molecular basis of somatic adaptability provides a sophisticated level of understanding of the connection between physiological change and genetic change. Ultimately many processes boil down to an increase or decrease in activity, so both environmental and genetic means can produce the same outcome in different ways.[2]

Our examples of somatic adaptations and their modification in evolution show the ease with which organisms can move between environmental response and evolutionary (genetic) modification. We will consider examples from insect castes, temperature-dependent sex determination in reptiles, and the role of hemoglobin in regulating oxygen levels. In all three cases a physiological mechanism generates alternative states that are elicited by external signals.

For insect castes, the queen and worker bees represent alternative states of the same genome. They are induced by chemical signals relayed during growth of the juvenile.

Sex determination in reptiles by ambient conditions is widespread. Here the alternative states are male or female, with no intermediates. Selection is done by the temperature at which the eggs develop.

In the blood of all vertebrate species, oxygen is transported by hemoglobin, a carrier protein. Hemoglobin has two interchangeable states with different oxygen affinities. In different tissues or environmental conditions, the exact affinity of the hemoglobin population for oxygen is changed by small molecules that shift the mix of the two states, whereas the affinities of the states themselves are unchanged.

To a greater or lesser degree, we comprehend the nature of these alternative phenotypes and the dichotomous decisions for reaching them. It is evident in all three cases that evolution has exploited this physiological plasticity to craft new phenotypes by substituting a genetic change for an environmental one. By understanding the nature of the switch, we can appreciate how easy it is to transform a physiological change into a more stable genetic one.

Schmalhausen and the Baldwin Effect

With the aim of finding a place for somatic adaptation in evolution, James Mark Baldwin (1861–1934) and others proposed the hypothesis of organic selection, which drew on Lamarck and Darwin without conflating them. Baldwin was an early experimental psychologist at a time when psychology was separating from philosophy as a field of study. Behavior had all the elements of somatic adaptation, and Baldwin proposed that animals have broad ranges of somatic adaptability enabling them to tolerate environmental change when their niche alters or when they enter new niches.

In the unfamiliar environment, the organism is stressed; hence it continues to utilize its adaptive mechanisms. It is not fully adapted, though viable enough to reproduce at least minimally. In subsequent generations, heritable changes arise in a few members of the population that fix the somatic adaptation and remove stress; these members are then selected for their increased reproductive fitness. In this way a heritable internal stimulus has replaced the external one in maintaining the adaptation. Thereafter the organism expresses it even when removed from the environmental stimulus.[3]

This hypothesis differs in several respects from the neo-Darwinian view, although to make the comparison we must translate some of Baldwin's pre-Mendelian language into modern terms. According to Baldwin, mutant variants would not have to precede selection: instead, mutation could follow selection. In the new condition, somatic adaptation would suffice at first for the survival of many members of the population, not just rare variants. Mutants and in particular new gene combinations would in time arise in a somatically adapted population. They would be selected as they stabilized, refined, and extended the somatic adaptation.

This phase of the process is simply Darwinian variation and selection. The somatic adaptation becomes a heritable evolutionary adaptation, persisting even under nonstress conditions when these arise. For Baldwin, mutation might do more than stabilize the somatic adaptation; it could stabilize it at a more optimal state.

Baldwin's proposal did not violate the Darwinian variation-selection principles, yet it gave prominence to individual somatic adaptability. Without that preexisting adaptability, the new selective conditions might extinguish the organism before it had a chance to adapt genetically. Baldwin avoided Lamarck's inheritance of acquired characteristics, which were somatic adaptations, because Baldwin required independent heritable genetic change. That took the form of new genetic combinations from the genetic variability in the population or new mutations to stabilize and fully express the adaptation. The main implication is that the complex phenotypic variation is not created from nothing, but rather from preexisting processes and components of the organism's somatically adaptable phenotype, whereas mutations merely stabilize and extend what is already there. Thus, the mutations need not be creative and numerous, a proposal that of course greatly reduces the difficulty in generating phenotypic change.

Ivan Ivanovich Schmalhausen (1884–1963) expanded the Baldwin proposals in his book *Factors of Evolution*, completed in Moscow in 1943 in the depths of World War II while the city was under Nazi siege. Schmalhausen was a leading figure in Russian biology before the war and until 1948. At that point Fyodor Lysenko, the notorious Stalinist scourge of Russian genetics, brought him to trial for being a "Weismannist-Mendelist-Morganist idealist."[4]

Schmalhausen began with the concept of the "norm of reaction" of an organism, that is, the range of phenotypes expressed when the organism reacts to various environmental conditions (temperature, humidity, crowding, kind of food). This reaction norm has two components. When the organism is stressed, some responses confer adaptive benefit; other responses are nonadaptive to environmental stresses. These morphoses (so named by Schmalhausen) are changes in the organism under stressful conditions, but they do not help the organism accommodate to that stress. For the example of *Drosophila*, Schmalhausen cites morphoses that include changes in the anatomy under conditions of increasing temperature, such as enlargement of the eye. The larger eye may give the fly no relief from heat, but may improve its getting around in dim light, an unrelated selective condition.

Taking together these two kinds of phenotypic changes, we would only know the organism's all-encompassing norm of reaction after it had been stressed in all combinations of conditions and durations of exposure. The span of responses would encompass the entire range of phenotypes the organism could generate from its single genotype; it would reveal the total latent phenotypic variation within the organism that could be generated without new genetic variation. This is certainly a broadening of Baldwin's view, including as it does not only the evoked developmental, physiological, and behavioral adaptations but the nonadaptive morphoses as well.

The adaptive reactions are more easily discussed, for they are used in the way Baldwin foresaw. When the organism enters or finds itself in a different environment, it adapts somatically to the extent it can. It survives, though stressed and perhaps marginally reproductive. Then heritable variants arise in the population that extend and stabilize the phenotypic adaptation, and they are selected. This is the Baldwin effect restated in more modern terms. After the stabilizing mutations, the organism is more reproductively fit and presumably less stressed. The change has been stabilized by internal heritable agencies rather than by external nonheritable ones, and the trait is produced each generation as part of the animal's embryonic development. A new norm of reaction has been generated, and bit by bit the organism can adapt to new circumstances.

The evolutionary significance of nonadaptive "morphoses" in the generation of phenotypic variation is harder to explain, except when the morphosis is fortuitously adaptive for some other selective condition. That is, one condition provokes or unmasks the morphosis, but a coincident, separate condition selects it.

In both the immediate adaptive and nonadaptive responses to environment, phenotypic change does not depend on new mutations or genetic variation. In the stabilization phase, the genetic variation may come from reassortment of existing variation in the population rather than from new mutations. The components and processes needed to produce the initial phenotypic variation are already there, without genotypic change. Before the organism meets the selective condition, its response is already encoded in its genome. Only "small" regulatory changes deriving from genetic variation in the population are needed to stabilize the change, bringing it under internal control. If the organism's "envelope of reaction" is vast, the organism as a whole can be described as a great exploratory system with many possible outcomes, from which random genetic change stabilizes a particular outcome relevant to the selective conditions. Thus, Schmalhausen simplified the thinking about phenotypic novelty by saying that it is within and we do not see it; it does not have to be created anew.

Experiments on the Baldwin Effect

At about this same time, Conrad Waddington in England began to think along similar lines and arrive at similar conclusions. His term for stabilizing selection was *genetic assimilation*, a term still in use, and he added two refinements as the result of experiments he performed with *Drosophila*. For example, he exposed fly populations to high salt in their food, to evoke their somatic adaptability toward salt; he then selected for increased salt tolerance. Or he exposed flies to ether during embryogenesis to provoke the later development of an extra pair of wings (four wings rather than two), a morphosis, after which he himself selected for flies with this outcome. (Extra wings seemed in no way to protect the fly against ether.) Or he shocked pupae at high temperature, which blocked the later development of cross-veins on the wings, after which he also selected for flies with this morphosis, as illustrated in

Figure 12. In all cases, he repeated the treatments and selections for 20–25 generations.[5]

First, he found, the initial population was usually quite heterogeneous in its response to the treatment, because of preexisting genetic variation and prior environmental conditioning. And second, much of stabilization, or genetic assimilation, was afforded by old genetic variation already in the population and brought into new combinations during successive matings, not by new mutations. The success of assimilation therefore depended on the genetic variation available in the population. With genetically heterogeneous populations, Waddington could, by the end of 20–25 generations of selection, reliably obtain populations showing the phenotypic novelty at high frequency, now independent of the special environmental conditions of treatment. With genetically homogeneous (inbred) populations, the effect did not appear.

Susan Lindquist is a geneticist who in the 1990s studied the resistance of organisms to heat. She extended the Waddington experiments to discern how heat unmasks cryptic phenotypic and genotypic variation. Excessive heat, like other stress conditions, causes most proteins of the cell to unfold and lose activity. The organism produces several kinds of special heat shock proteins (Hsp), or chaperone proteins as they are called, that guide the refolding of unfolded proteins back into their active form, thereby mediating recovery from heat. Hsp90 is one of these proteins.

As it turns out, even without heat, the Hsp90 protein is continuously important for folding newly made proteins correctly, especially large proteins of signaling pathways. When the organism is heated, the Hsp90 is recruited to refold damaged proteins, but there is not enough Hsp90 for folding the new proteins most in need of chaperone assistance. Aberrant phenotypes emerge—not just the cross-veinless kind pursued by Waddington but a wide range of others. Any of them could be selected by the researcher, and after some cycles of heating and selection would become stabilized. The spectrum of altered phenotypes differs in various stocks of flies, showing that genotypic diversity exists relative to which proteins are most dependent on chaperones.[6]

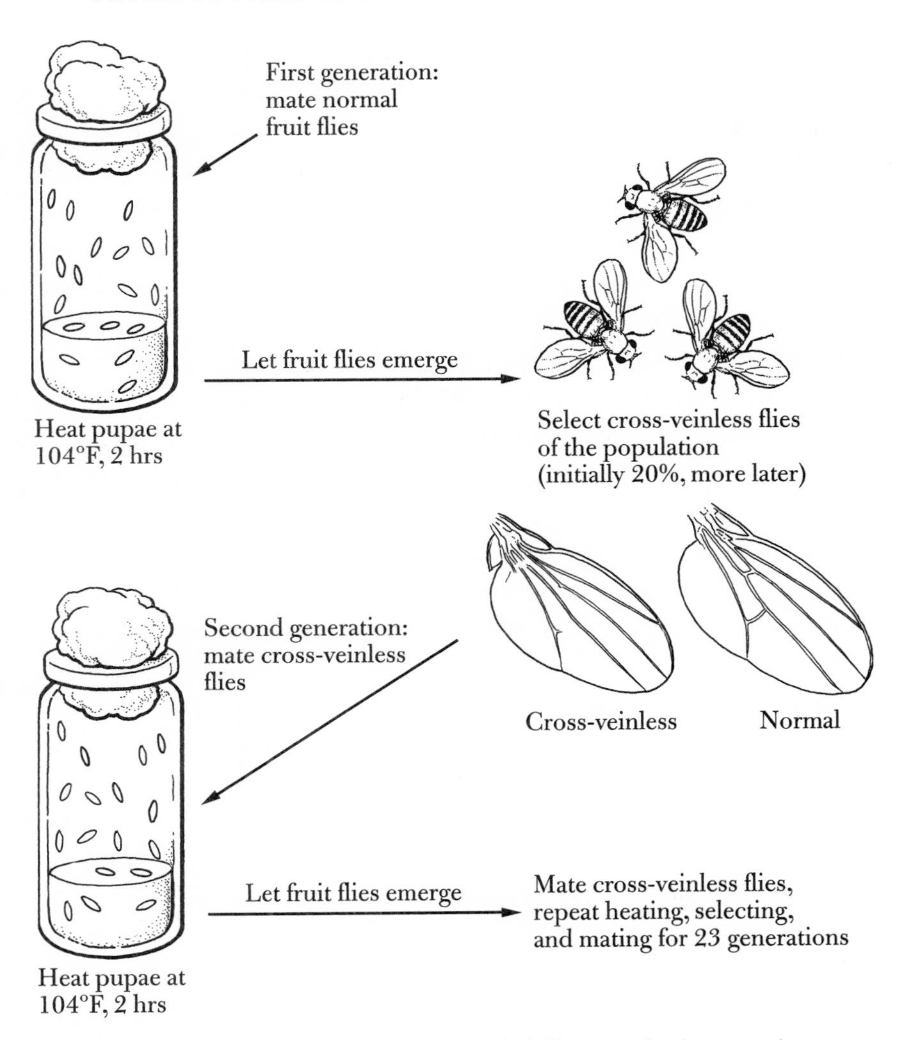

Figure 12 The Waddington experiment. A deliberate selection experiment demonstrating genetic assimilation after exposure to heat. By the twentieth generation, some flies developed cross-veinless wings even without heating. Note that the cross-veinless trait confers no benefit against heat.

Lindquist and her colleagues tested not just *Drosophila* but also *Arabidopsis,* a small flowering plant. They lowered Hsp90 activity in several ways, by heat (as described above), by mutation of Hsp90 itself, and by exposure of the organisms to a chemical agent that inhibits hsp90 specifically. Many alterations of morphology were seen and, if desired, these could be put through stabilizing selection so that they would continue to form even after Hsp90 was restored to full activity. The variety was great: different altered parts and different degrees of alteration. Lindquist refers to hsp90 as a "capacitor for phenotypic variation." (Capacitors in electrical circuits accumulate a reservoir of charge and release the charge when there is a change in the circuit.)[7]

Several evolutionary biologists, including Mary Jane West-Eberhard, along with Carl Schlichting and Massimo Pigliucci, emphasized the possible broad applicability of this adaptation-assimilation hypothesis in explaining the origin of complex phenotypic variations. In West-Eberhard's view, evolution of a novelty proceeds by four steps. In the first, called trait origin, an environmental change or a genetic change affects a preexisting responsive process, causing a change of phenotype (often a reorganization). At this initial step, she regards environmental stimuli as apt to be more important to evolution than genetic variation. The traits may or may not be adaptive; if they are not, they resemble Schmalhausen's morphoses or Waddington's and Lindquist's temperature-evoked changes.

In the second phase, the organism adapts or accommodates to its changed phenotype by compensating in part for the perturbed condition by using what we would say are its highly adaptive core processes.

In the third phase, recurrence, a subset of the population continues to express the trait, perhaps owing to the continued environmental stimulus.

In the final phase, genetic accommodation, selection drives gene frequency changes that increase fitness and heritability, although the phenotypic change is not necessarily ever completely under genetic control. While having a heritable component, it could retain an environmental dependence.

Thus this model, like Schmalhausen's, has a phenotypic accommodation phase followed by a genotypic accommodation phase. Most elements of a phenotypic novelty would not be new, and the role of mutations would be to provide small, heritable regulatory modulations rather than to create major innovations.[8]

Leading evolutionary biologists, including creators of the Modern Synthesis such as Ernst Mayr and George Simpson, were not impressed by adaptation-assimilation ideas. Some critics, missing the point, said that somatic adaptation is not heritable and hence is irrelevant to evolution, to which physiologists said the capacity for generating a broad range of somatic adaptations is as heritable as anything else. Other scientists acknowledged that the Baldwin effect "probably has occurred, but there is singularly little concrete ground for the view that it is a frequent and important element in adaptation."[9]

Simpson, the leading paleontologist of the Modern Synthesis, doubted that most traits could be regarded as stabilized somatic adaptations. He was most interested in anatomical changes, not physiological and behavioral alterations. Overheating animals might indeed create a few anomalies, such as larger eyes in fruit flies, but it was hard to imagine that it could elicit major anatomical innovations like jaws or wings. The somatic changes seemed quantitative, not qualitative—strengthening and weakening existing traits, not inventing new ones. In the middle of the twentieth century, evolutionary biologists tended to conclude that the Baldwin effect, if it exists, is of only minor relevance to evolution. Yet in a more modern molecular context, we believe that somatic adaptation, when applied to the conserved core processes, can help us resolve vulnerabilities in evolutionary theory about the origin of novelty.

A second reservation was that the adaptation-assimilation ideas, if correct, would have required the organism to hold much of future evolutionary adaptation within itself; it was hard to visualize how this broad potentiality would have been selected previously in evolution. The Baldwin effect seemed suitable for exploration of small deviations from the existing phenotype but not for radical new experiments. The organism might have a reaction norm for temperature, which would allow rapid evolutionary change within a certain range, but after exceeding that range, multiple characters would seemingly have to change at once. For physiological ranges that the organism might never see, such as the invention of new structures like eyes and wings, the organism in all likelihood would not contain within itself a plasticity extending to phenotypes it had never explored. Such reservations might not extend to morphoses, since these had not been previously selected for adaptability.

Plasticity in Development

The capacity of an organism of a single genotype to generate two or more phenotypes by alternative developmental paths is an example of the adaptive norm of animals to which Schmalhausen drew attention. Developmental plasticities differ from physiological adaptations in that after a critical time in the animal's development they are irreversible. The alternative phenotypes can be distinguished by their morphology, physiology, or behavior. Organisms with alternative phenotypes are called polyphenic.

There are two principal kinds of polyphenism, sequential and alternative. Taking sequential phenotypes first, the cases include animals with complex life cycles of two or more different developmental stages (such as larva, juvenile, and adult). Most animal phyla inhabit the ocean, and most pass through strikingly different developmental stages. The larva might feed in the plankton-rich surface layer of the ocean, whereas the adult might live in the mud, sand, and rocks near the shore. In the case of ascidians (sea squirts), the larval and adult forms look so different that they were thought to be members of different phyla until the late 1800s, when the development of the larva to the adult was followed. Some parasitic flatworms (trematodes) have five or six successive forms, each highly specialized for different lifestyles in different hosts. Of course, for terrestrial animals, we are familiar with the vegetarian wall-eyed swimming tadpole as the amphibian larval phenotype and the carnivorous four-legged frog as the adult phenotype—or the wingless, legless caterpillar and the adult butterfly.

When stages of a life cycle differ dramatically, they are connected by a drastic metamorphosis. In the metamorphosing caterpillar, most larval tissues are destroyed and replaced by newly developed adult cells. The transition from one stage to the next is usually dependent on external conditions, although it must be appreciated that this dependence is evolved and selected. Two hormones control insect metamorphosis: ecdysone, a steroid-like hormone, and juvenile hormone, a close relative of vitamin A. Though made by the insect, their times of synthesis and effectiveness are subject to aspects of nutrition, light, and temperature.

Unlike hormones, such as insulin, which are involved in maintaining a stable internal environment ecdysone and juvenile hormone do not maintain the original state; rather they globally release the internal means to propel the organism to a new state. In butterflies, there are specific times in the last larval period when the full-grown caterpillar responds to juvenile hormone by turning off genes appropriate for the larva and turning on genes appropriate for the pupa. These effects, though far-ranging, are mechanistically no different than the common process of gene regulation that occurs in all cells of our body. The timing of the response to juvenile hormone is itself regulated by periodic pulses of ecdysone. Different tissues respond differently to juvenile hormone and ecdysone, various cells responding to one or the other, both, or neither. Where there has been a response to external conditions, it is the organism that specifies the response, the readiness to respond, and the specificity for responding to a particular environmental agent. Thus, the same genome can be read differently at different times to drastically alter the phenotype. The timing of these events can be linked to the external environment or can be driven by purely internal means.

Sequential phenotypes have in some cases provided the phenotypic variety for the founding of new races or species. Salamander species exist in which the "adult" is basically a large larva that has gained sexual maturity. It retains larval traits such as a finned swimming tail, large external gills, and an aquatic lifestyle. Metamorphosis from the larva to adult, which is normally thyroid hormone dependent, is largely forgone.

The lake-dwelling axolotl of the Mexican highlands is a well-known example. When thyroid hormone is experimentally provided, the animal completes metamorphosis and comes to resemble a related land-dwelling species (of Texas). Plausibly, the axolotl is derived from a metamorphosing ancestor, in which sexual development was prematurely completed in the larva and thyroid hormone production was lost by a heritable defect. It has been argued that the Mexican highlands are cool and deficient in iodide, two conditions known to hinder or block metamorphosis. The axolotl may have overcome the conditions by omitting metamorphosis altogether. Its apparent morphological novelty as an adult is the retention of larval features already present in the ancestor, not the origination of new features. This kind of persistence of the juvenile form is common in salamanders.

In another case, the landlocked salmon of freshwater lakes is thought to be an arrested alternative phenotype of an ancestor that moved between saltwater feeding grounds and freshwater spawning grounds. The landlocked adult resembles a "parr" form of salmon, which is a dark, bottom-dwelling juvenile that in the ancestor would have undergone a thyroid-hormone–triggered metamorphosis to a silvery migratory adult.

Whereas these examples involve the retention of larval or juvenile traits, other cases involve the loss of the larval stage and a direct development of the embryo to the adult. In sea urchins, most species produce small eggs that develop to bilateral plankton-feeding larvae that, after much growth, metamorphose to a penta-radial adult, a radical transformation of body organization. In every family of sea urchins are species that have evolved a direct form of development that omits the larval stage. Their eggs are larger and contain more yolk. The large embryo develops directly into a small adult. The larval feeding stage has been omitted, and the bilateral development of the embryo has been modified—nearly skipped—to yield directly a penta-radial outcome. If one unknowingly compared what hatched from the eggs of closely related direct and indirect developing species, one would say the difference is enormous; but in fact the one species represents only an omission of part of the adaptive norm of reaction of the other. Little evolutionary novelty is involved. Novelty has been generated, but it has been generated by omission rather than by creating new processes.

In some ways the more interesting forms of developmental plasticity occur when the organism has alternative adult phenotypes developed in accordance with environmental or social conditions, the second kind of polyphenism. The irreversibility of this polyphenism is due to an environmentally dependent branch point, controlled by a sensitive switch, at one episode of development. When this decision has been made, it cannot be repeated or undone.

Social insects such as ants, wasps, bees, and termites offer dramatic examples of alternative adult phenotypes. Honeybees are a favorite experimental model of phenotypic plasticity. When the dominant queen emerges from the pupa slightly ahead of contending queens, she takes over the nest. The older mother queen leaves, taking a host of worker bees with her. These are her sterile diploid sisters, sharing three quarters of her genetic information. The fertile queen and sterile workers are obviously alternative phenotypes of the same genotype.

Compared to the workers, the queen is larger but has smaller eyes, reduced mouthparts, shorter antennae, a smaller brain, no pollen-collecting combs, rakes, or baskets on the leg, and poorly developed glands of the sort used by workers to build the waxy hive and to feed larvae with royal jelly or worker jelly (illustrated in Figure 13). However, the queen develops very large ovaries and specialized glands producing "queen substance," which controls the behavior of the workers. The queen is the reproductive alternative phenotype, fed by the workers and producing over a thousand eggs a day. She is figuratively the germ line of the colony. The workers, who are sterile because their ovaries remain undeveloped, collect food, do the waggle dance to inform other workers of the location of flowers, deposit honey, build the cells of the hive, supervise the deposition of eggs, feed the larvae and nurse workers, and air-condition and clean the hive. They are the soma of the colony. Clearly, the queen and the worker are very different phenotypes.

How do they develop one way or the other? Both come from the same kind of diploid egg at the start, as shown by experiments trans-

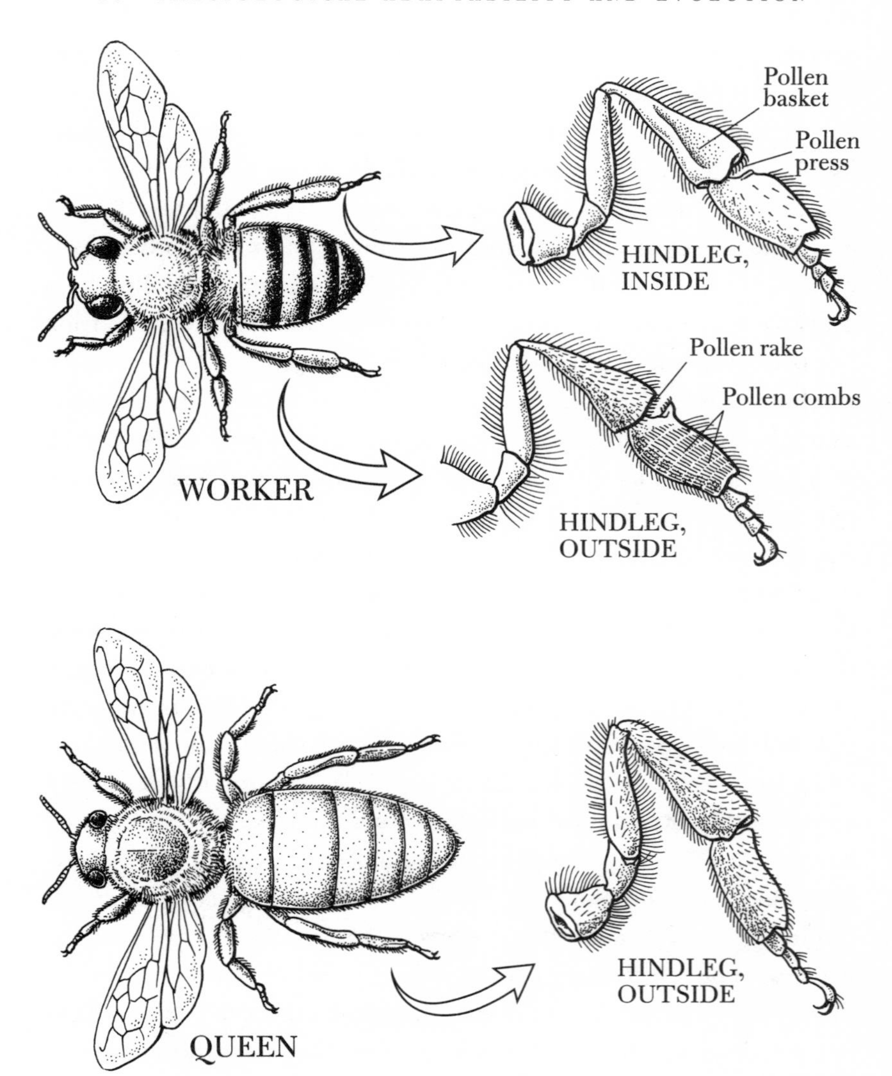

Figure 13 Castes of honeybees. The worker and queen are sisters raised under different conditions as larvae in the hive. The legs develop differently, with only those of the worker specialized for pollen collecting. The inside of the queen's hindleg looks the same as the outside.

ferring eggs between royal cells and worker cells—whatever egg is in the royal cell becomes a queen. Workers construct royal cells at a time when the queen mother produces insufficient queen substance to inhibit them from doing so, and the workers start preparing royal jelly and feeding it to larvae in these cells. Royal jelly is a nutritious substance, which in honeybees includes high concentrations of vitellogenin, a protein found in both vertebrate and invertebrate egg yolk. By the third day of larval life, the prospective queen differs from the prospective worker in its development, as shown by the fact that a change of diet to worker jelly or a change to a worker cell no longer alters the outcome. Queens develop faster and emerge from the pupa after only 16 days, whereas workers emerge at 21 days. The first queen out may kill her tardy sister queens and take over the hive (the mother queen has already left), or she may leave in a swarm.

It is during the larval period that the choice is made to develop to either the worker or the queen. Strong evidence exists that royal jelly and worker jelly differ in controlling the level of a hormone needed during the brief time of larval development to the queen. Royal jelly raises the level of juvenile hormone during this critical period; in fact, topical application of juvenile hormone can induce the formation of a queen. Research is just beginning on how the larva responds to the royal and worker jellies to generate the different morphs. It is known that several genes are differently expressed in the worker and the queen. Workers, though smaller, are in fact more complex in their anatomy and physiology than queens, which are basically egg-producing factories with many rudimentary body parts.[10]

Polyphenism is not limited to insects. A predatory species of cichlid fish (*Cichlasoma managuensa*) can develop either blunt jaws for biting prey or pointed jaws for sucking in prey. If newly hatched fish are raised in the laboratory under two feeding conditions, their jaw development is different. The young start with small, blunt jaws; if they feed on flake food for 16.5 months, they develop into adults with full-sized, blunt jaws. If they instead feed on brine shrimp for the same period, they develop into adults with large pointed jaws. The difference in jaw structure is apparent by 8.5 months. If the blunt-jawed fish eating flake food are switched to brine shrimp at this time, they can still develop pointed jaws, but if switched soon thereafter, they cannot. Jawbone development before 8.5 months probably responds to the dynamic load of feeding.[11]

Cichlid fish, as a worldwide group, are unusual in having a second set of jaws in the back of the mouth: the pharyngeal jaws, shaped from a throat bone. The availability of two jaws has allowed a division of usage, the front pair for ingesting the food and the second pair for chewing it. Humans, on the other hand, demand both functions from one pair of jaws, and each function probably limits the specialization of the other. This plasticity of development concerns the front jaw of this particular species of cichlid. In another, the development of the back jaw also shows developmental plasticity. It becomes wider and fills with crushing teeth when the diet is rich in mollusks rather than insects.

Overall, the phenotypic plasticity of jaws and the presence of independently specialized jaws in cichlids may explain their position as one of the most species-rich groups of animals, with as many as five hundred species in just one African lake. Some of these species are thought to have arisen by the mutational stabilization of one or the other alternative phenotype of an ancestor possessing rich developmental adaptability.[12]

Plasticity and fixation may underlie much evolutionary change. The different developmental alternative phenotypes of an organism's complex life cycle, and the alternative adult phenotypes, are aspects of the total phenotypic plasticity of the organism (the whole adaptive norm of reaction of Schmalhausen). Evolutionary specializations, made heritable by new genetic variation, are but stabilizations of certain already-available states, in the mode envisioned by Baldwin, Schmalhausen, Waddington, and West-Eberhard.[13]

Environmental and Chromosomal Sex Determination

In the course of evolution, some processes move easily from environmental control to genetic control and back again. It may seem odd to think of sex determination as a response to the environment, but it is one of the clearest examples of the interchangeability of physiological and genetic control. In many organisms, fish and reptiles included, the ratio of males to females may deviate far from unity depending on environmental conditions such as temperature or social interactions. Summarizing the knowledge in 1900, E. B. Wilson (later, ironically, one of the discoverers of sex chromosomes) wrote: "Sex as such is not inherited. What is inherited is the capacity to develop into either male or female, the actual result being determined by the combined effect of conditions external to the primordial germ-cell." The discovery in 1905 of the genetic basis of sex is considered one of the great triumphs of early-twentieth-century biology. Nettie Stevens of Bryn Mawr College provided convincing evidence for the control of sex by the balance of X and Y chromosomes, based on her studies in over 50 species of beetles. She found that females always possessed two X chromosomes; males usually had one X and a smaller Y chromosome, although some species lacked a Y altogether. She and Wilson independently published monumental papers concluding that chromosomes determine sex. As we noted in Chapter 1, T. H. Morgan's first great achievement in genetics was his analysis of a white-eyed male, where eye color was linked to maleness. It gave incontrovertible functional proof in fruit flies (later in many other animals) that females carried two copies of the X chromosome, whereas males carried one. In contrast, organisms with environmental sex determination showed no chromosome difference between males and females.[14]

It is surprising that the mechanism for sex determination is not as universal as the processes of meiosis and fertilization on which sexual reproduction depends. In crocodiles and many species of lizards and turtles, sex is determined in an extremely narrow range by the ambient temperature at which the egg develops. For example, American alligator eggs incubated at 86° F (30° C) produce 100 percent females, whereas at 91° F (33° C) they produce 100 percent males. The embryonic animals are sensitive to temperature during a particular week of their nine-week period of development, as the sex organs begin to take on characteristics of either the testis or the ovary. Before this time, the developing sex organs look the same in all individuals and can become *either* testis or ovary. The organ is called the indifferent gonad at this stage. At 91° F (33° C) in the critical week, large cells proliferate and surround the germ cells. They will form the Sertoli cells, which play a critical role in the development of the testis and spermatozoa; a male results. At 86° F (30° C) in the critical week, the germ cells proliferate and form clusters. The large cells fail to proliferate and disappear; an ovary develops and a female results, as illustrated in Figure 14.[15]

What is the temperature-dependent step of sex determination of reptiles, recognizing of course that the early development of the gonad is the same in both sexes and is temperature independent? And what has replaced the temperature-dependent step (if that is the right way of thinking about it) in mammals and birds, all of which have chromosomal sex determination? How could testis and ovary development look initially so similar in mammals and reptiles, and at the same time be under such different controls?

The key to understanding the physiology is a circuit with two stable states (male and female), a circuit that can flip-flop between the two. At intermediate temperatures, intermediate proportions of males and females are produced, not intersexes or hermaphrodites. One clue is that in the critical period of development, alligator eggs at the temperature normally producing males can be entirely switched to females by exposing them to estradiol, the female sex hormone. Similarly, under conditions where females are normally produced, chemically inhibiting estradiol synthesis leads to the development entirely of males.

One may infer that if estradiol production were itself temperature sensitive, it could serve as the trigger for environmentally controlled sex determination. Carefully poised switches would be easy to control genetically as well as environmentally and could bring an entire suite of activities under the control of mutation, facilitating rapid evolutionary changes. To appreciate how switches could be thrown both genetically and environmentally, we need a molecular description of the process.

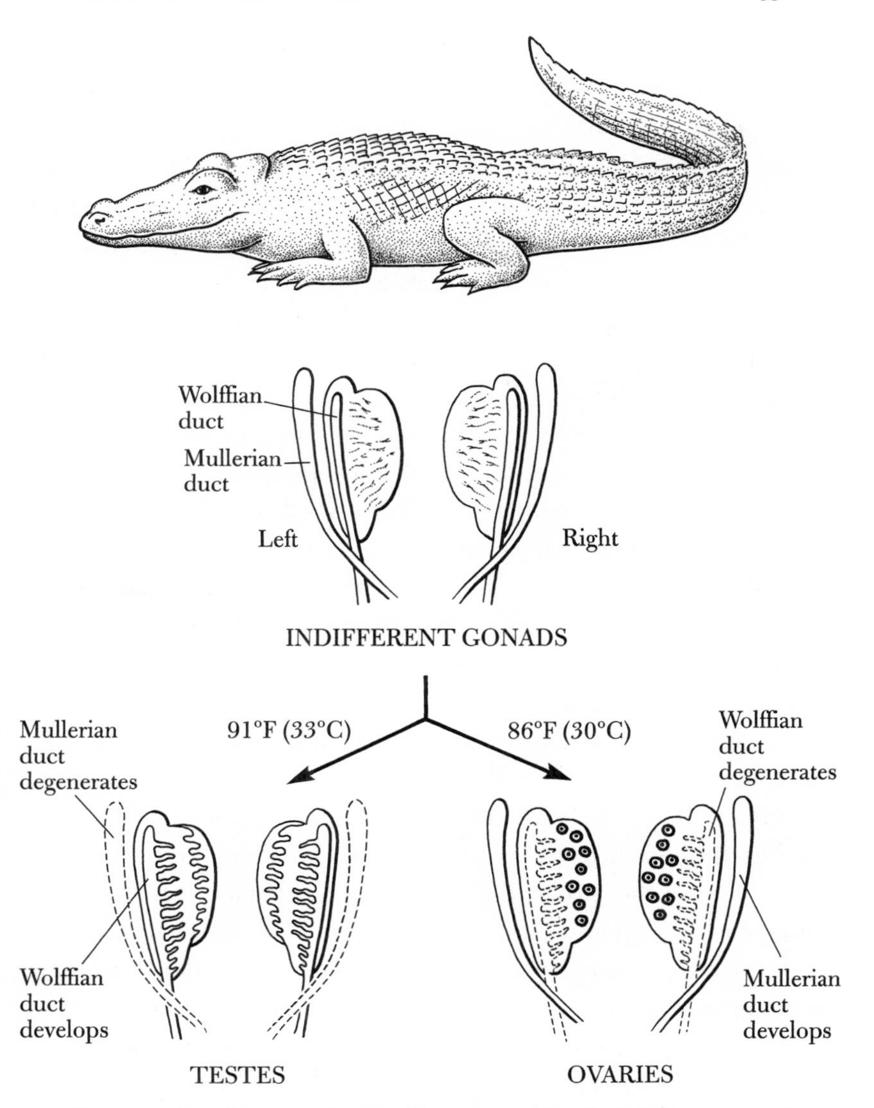

Figure 14 Sex determination in alligators. At first the indifferent gonad develops with parts for both the testis and the ovary. Later it destroys some parts and further develops others—for either testis or ovary, but not both. The Wolffian duct becomes the epididymus and vas deferens in males. The Mullerian duct becomes the oviduct and cervix in females. The temperature at a critical time decides the direction.

Such information is now just appearing for sex determination in turtles. Experiments on the red-eared slider turtle argue that control of the pathway for making estrogen from another steroid hormone (a testosterone-like steroid) is in fact the trigger. In this turtle, males are produced at lower temperatures 78° F (26° C) and females at higher, 88° F (31° C), the opposite of the alligator. In a flip-flop circuit, not unlike a thermostat that would gratify any engineer, a small difference in the level of a regulator of estrogen synthesis can be amplified into one of two states, a high-estrogen state (female development) or a low-estrogen state (male development).

The key to the circuit is a gene called SF-1, whose encoded protein regulates the expression of other genes encoding enzymes for making both testosterone and estrogen. At higher levels of SF-1 protein (low temperature), enzymes are made that synthesize testosterone as the major product, whereas very little estrogen is produced. Males result. At lower levels of SF-1 protein (high temperature), the enzyme that converts testosterone to estrogen is made, and estrogen is the major product. Females result.

An added feature of the circuit ensures that no intermediate states are generated that would lead to intersexes. When estrogen begins to build up, it feeds back and inhibits testosterone production. The result is a bistable switch driven one way or the other by the temperature dependence of the production of SF-1 protein. Since virtually everything in biology is temperature dependent, it is easy to see how various forms of temperature regulation and genetic regulation may have evolved many times in different species of reptiles and fish.[16]

Some fish have such labile sex determination that they can change sex depending on social circumstances, rather than temperature. In the bluehead wrasse, loss of the dominant male in the population allows the largest female rapidly to adopt male behavior, later to assume male coloration, then to convert the gonad from an ovary into a testis. The behavioral changes occur within hours, independently of the gonad. They are contingent on the production of a hormone secreted by the brain.[17]

In vertebrates, the genetic process of sex determination falls into two classes. In mammals, the female has two identical sex chromosomes (XX), whereas the male has one X and one small Y chromosome (XY). In birds and amphibians, the male has two identical sex chromosomes (ZZ), whereas the female has two different ones (ZW). How did these two varieties of chromosomal sex determination arise, and how are they related to the means of environmental sex determination? Most important, what do the answers to these questions imply about how processes of somatic variation can facilitate evolutionary change? The underlying mechanism of sex determination is the same in all vertebrates. All have the same enzymes under the control of conserved factors like the SF-1 protein. The original vertebrate ancestor, like the present-day turtle, most likely employed some form of environmental sex determination and had no sex chromosomes.

Genes, because they are under selection, are rather stable in evolution; chromosomes are not. Genes may be copied and moved from chromosome to chromosome. A particularly brisk traffic of genes occurs on and off the X chromosome in mammals. One reasonable scenario for the evolution of genetic sex determination holds that one member of a pair of chromosomes, both of which initially carry the same genes for various components of sex determination, began to degenerate, losing more and more genes to other chromosomes. It is now evident that the Y chromosome in humans is the paltry residue of an ancient X chromosome that began to disintegrate millions of years ago. The Y chromosome still contains remnants of X sequences. In the process of degeneration, the diminishing X chromosome (on its way to becoming a modern Y) would cause an imbalance of sex-determining factors. In the case of mammals, it (now called the Y chromosome) would have retained male-determining genes, whereas in the case of amphibia and birds, it (now called the W chromosome) would retain female-determining genes. At this point, sex determination would be genetically based because the Y chromosome would be necessary for "maleness" in mammals and the W chromosome for "femaleness" in birds.[18]

Where will this all lead? The degeneration of the X chromosome, which produced the vestigial Y, need not stop until the Y itself is completely destroyed. We may have the answer in a mammal called the mole vole (genus *Ellobius*), which appears to have no Y chromosome at all. There is no reason why the sex-determining activity has to remain on the degenerating Y chromosome; it could migrate to another chromosome, producing a new incipient sex chromosome. When that chromosome degenerates, it will become a new Y chromosome.[19]

The unexpectedly wild evolutionary history of sex determination tells us that both evolution and physiology can play upon the same poised system. Such examples readily distinguish regulatory events from core processes. The core process in vertebrate sex determination is the control of estrogen production through a bistable circuit involving the regulatory factor SF-1. The core process is conserved. What is highly diversified is the means that animals use to tip the balance of the core process toward the male outcome (low estradiol) or toward the female outcome (high estradiol).

Regulatory tipping of the heritable genetic kind includes the imbalance of genes in the XX versus XY alternatives, but also includes in other animals a number of environmental factors. Since all enzymatic reactions are temperature dependent, responsiveness to temperature is naturally available to the system. The impressive machinery is the core system of estradiol production itself, poised to flip-flop into one of two modes, male or female, and to avoid an intersex intermediate. The triggers, which need generate only a slight bias, are easily evolved.

Hemoglobin: A Molecular Link Between Physiology and Evolution

Until the widespread investigations using the methods of molecular biology and biochemistry in the mid twentieth century, it was difficult to find evidence that the processes underlying somatic adaptability actually serve as a basis for evolutionary change (even though an environmental stimulus seems to be able to substitute for a genetic change). Such evidence requires a molecular understanding of the physiological or developmental process and a similar understanding of the critical evolutionary changes, so that they can be compared. Though most physiological and developmental processes are exceedingly complex, one physiological process stands out for its simplicity and offers a clear view of the relationship between somatic adaptability and evolutionary adaptability: the transport of oxygen by hemoglobin.

When William Harvey (1628) showed that the function of the heart is to pump blood through a closed circulatory system, he could offer no satisfactory explanation as to why blood passed through the lungs in copious quantities beyond the needs of nutrition. It was not until Joseph Priestley (1733–1804) and Karl Wilhelm Scheele (1742–1786) discovered oxygen, and until Antoine Lavoisier (1743-1794) demonstrated the consumption of oxygen in combustion and respiration, that living organisms were recognized to consume oxygen and excrete carbon dioxide. Blood contains more oxygen per unit volume than can be dissolved in water, and arterial blood contains more oxygen than venous blood. The excess capacity of blood to carry oxygen results from the binding of oxygen to an iron-containing pigment (heme, colored red as we know it) bound to a protein (globin).[20]

Almost all the complicated regulation and adaptability of oxygen delivery is incorporated in the relatively simple hemoglobin molecule and, as we shall see, important evolutionary adaptations are found there as well. Hemoglobin is an ancient molecule found in all life forms, from bacteria to plants and animals. In vertebrate animals hemoglobin is an aggregate of four protein molecules, two of the alpha globin kind and two of the beta globin kind. To each of the four globins is attached a heme group containing an iron atom. One oxygen molecule will bind to the iron atom of each. A hemoglobin molecule, the four-part aggregate, becomes saturated when each of the four irons is occupied by oxygen.

The function of hemoglobin is to load up oxygen in the lungs, transport it to the tissues, and unload it there; reciprocally, it loads up carbon dioxide in the tissues, transports it to the lungs, and unloads it. Although this process would seem to be simple (since there is more oxygen in the lungs and more carbon dioxide in the tissue), efficiency is important. An oxygen-binding protein without hemoglobin's special features might never be able to unload more than half of its oxygen during the circuit from lungs to tissues and back to lungs. The other half would be returned, unused. This in fact happens in the normal resting human.

What about the conditions of high oxygen demand when muscles are fully exerted? An optimal carrier, like hemoglobin, can unload nearly all its oxygen under these conditions. This two-fold boost in efficiency is of considerable selective importance during exertion. Imagine, for example, if the alternative were to gain the two-fold increase in oxygen delivery by doubling the size of the heart and blood vessels.

Hemoglobin is the best-known example of a protein that can exist in two conformations, two overall shapes of the protein, as opposed to only one. The four globin subunits are packed together in different arrangements in the two conformations, and the folding of each globin chain is slightly altered. One conformation of hemoglobin binds oxygen with high affinity; it is called the active or oxygen-loading state. The other conformation binds oxygen at five hundred fold lower affinity; it is called the inactive or oxygen-unloading state. The transition between the two states is all or none, that is, all four subunits of a hemoglobin molecule are either in the inactive state or in the active state, in concert, and the entire molecule changes rapidly from one conformation to the other, as shown in Figure 15.

This behavior holds the secret to hemoglobin's efficiency in oxygen delivery: it has a loading form and an unloading form, and can manifest one or the other depending on conditions (see Figure 16). When no oxygen is present, the inactive state is more stable, predominating ten thousand to one over the active state in the hemoglobin population. Oxygen binds to both states, but since it binds much more tightly to the active state, it hinders the return of that state to the inactive alternative. Thus, in the presence of oxygen, more members of the hemoglobin population are in the active state, and the population as a whole shows increased affinity for binding subsequent oxygen molecules. When two oxygens are bound per hemoglobin molecule, on average, the population has shifted so far that active and inactive

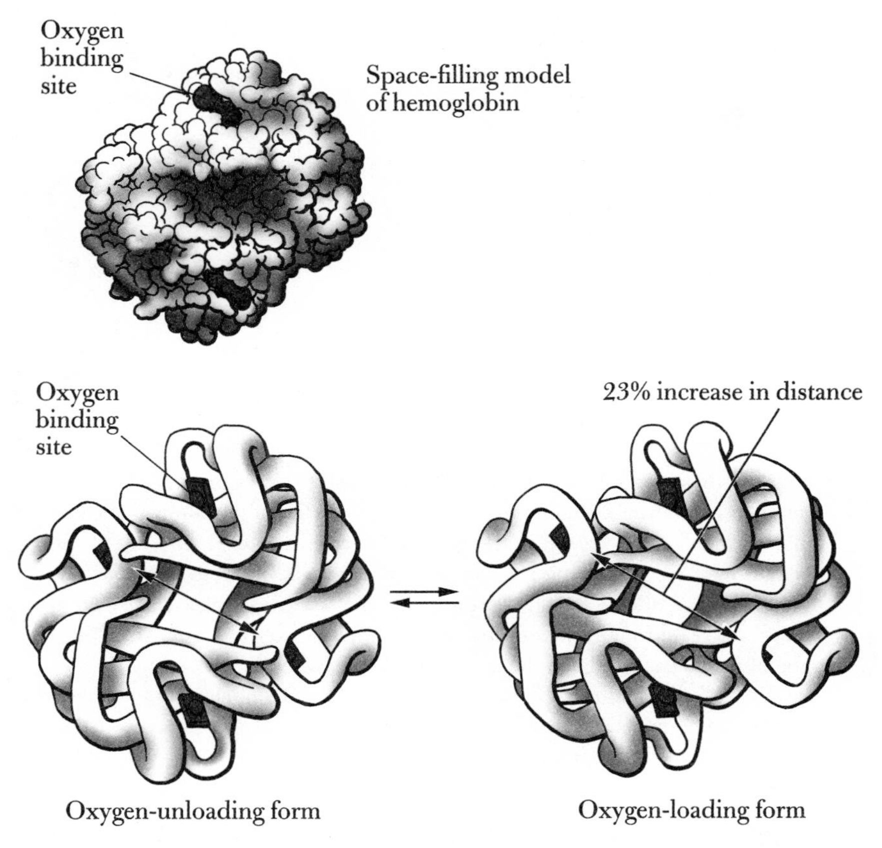

Figure 15 The two states of hemoglobin. *Above left*, the hemoglobin molecule, magnified ten million fold. The dark squares are the sites of oxygen binding. Only two of the four sites are visible. *Lower left and at right*, the hemoglobin molecule actually has two forms, shown by the exposed backbone. These forms interconvert rapidly and spontaneously. The form at the right loads oxygen well and the form at the left unloads it well.

states are equally present. As the fourth oxygen is bound, on average, the active state predominates by nearly ten thousand fold. The effect of this population shift is that oxygen loading is not gradual but precipitous, because the more oxygen a hemoglobin molecule binds, the more favored is its binding of the next oxygen.

The unloading process is also precipitous: as one oxygen is lost,

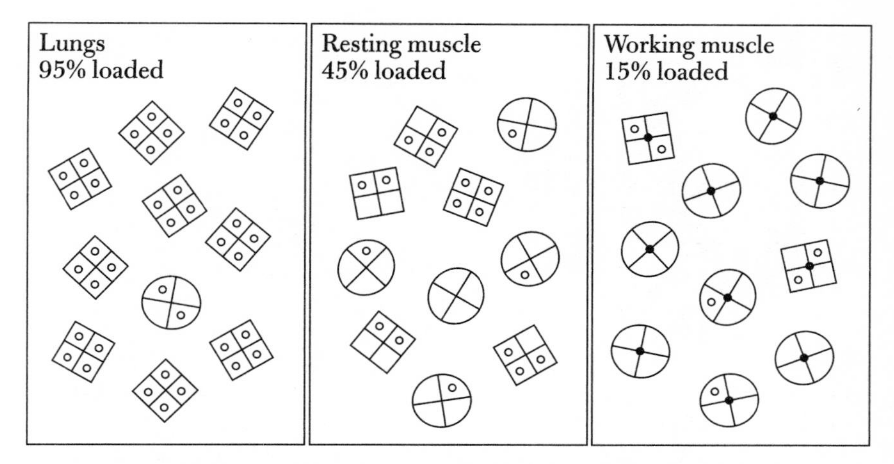

Figure 16 Responses of hemoglobin. *Left*, as blood passes through the lungs, most hemoglobin molecules are in the oxygen-loading form, shown as large squares. The bound oxygen is shown as small open circles. *Center*, as blood passes through resting muscles, about half the hemoglobin molecules flip to the unloading form, shown as large circles. Half the oxygen is released. *Right*, as blood passes through the muscles during extreme exertion, almost all the hemoglobin molecules flip to the unloading form, which is favored by the heat, acidity, and diphosphoglycerate (black dot on each hemoglobin molecule) present when muscles exert. Under these conditions, 80–90 percent of the oxygen is unloaded.

the more favored is the loss of the next one. At high oxygen levels in the capillaries of the lungs, hemoglobin is almost entirely in its active state and fully loaded with oxygen. At low oxygen levels, as occur in the tissues of the body during exertion, where oxygen is being rapidly and irreversibly consumed in energy production, hemoglobin is almost entirely in the inactive state and fully unloaded. It responds to oxygen absence by reducing its affinity for oxygen. It unloads it. The relative stability of the two states, and the relative affinities of the states for oxygen, are set within the protein molecule in such a way that for standard conditions (sea level, modest activity) hemoglobin almost fully loads up on oxygen in the lungs and dumps about half of it in the tissues.

We can imagine that a variety of conditions might perturb this equilibrium and thereby affect the loading-unloading behavior of hemoglobin. This is precisely what occurs, owing not only to external environmental changes but also to internal mutational changes in the protein. The environmental conditions are related to excessive oxygen demand (hypoxia), temperature, and acidity in the vicinity of hard-working tissues; the adjustments of hemoglobin are generally those that achieve maximum loading-unloading. It is a remarkably responsive protein, prone to regulation. When muscles are maximally active, the released heat and acidity favor the unloading conformation of hemoglobin. In response to these conditions, hemoglobin unloads 80–90 percent of its oxygen, a near-perfect transport agent.

Under unusual physiological conditions such as hypoxia, which we experience when we hike to high altitudes, the entire equilibrium of states, which is optimized for sea level, must be reset. The major effect of exposure to altitudes of above 12,000 feet (3,658 m)—two thirds the oxygen of sea level—is an increase in breathing rate. An elevated rate is maintained even after acclimation. Increased breathing leads to better oxygenation in the lungs. Although that improves the oxygen loading, it doesn't improve oxygen dumping in the tissues, and the whole process, despite the increased breathing rate, is inefficient. To reset the system to allow more efficient dumping under conditions of lower acidity, the red blood cell under hypoxic conditions makes a small molecule called 2,3 diphosphoglycerate, a by-product of its metabolism. This molecule binds to the inactive state of hemoglobin and tips the equilibrium toward the unloading form, to allow efficient dumping of the oxygen in the tissues.

The simple two-state equilibrium allows for regulation by both activators and inhibitors of oxygen binding. One activator of oxygen binding, as we have seen, is oxygen itself. The binding of oxygen on one of the four hemoglobin subunits facilitates the binding of a second oxygen molecule to a *different* subunit by shifting the population toward the active state, that is, the state with higher affinity for oxygen at all four sites. Inhibitors, such as hydrogen ions (acid) and diphosphoglycerate, inhibit by binding more strongly to the inactive state, shifting the population in that direction and all four site become simultaneously poor at binding and good at unloading (Figure 16).

Whereas mammals use diphosphoglycerate, other animals use different tricks to reset the oxygen balance under hypoxic conditions. Birds use a different sugar, inositol pentaphosphate, which like diphosphoglycerate binds to the inactive state and causes the oxygen to be dumped in the tissue. The lamprey has perhaps the simplest system. Lamprey hemoglobin exists in a two-state equilibrium where the active state is a single folded chain, a single subunit, of hemoglobin. The inactive state, a complex of two subunits, does not bind oxygen. Clearly, by binding only to the single subunit, oxygen shifts the population to the dissociated active state and increases oxygen loading. Acidity in the tissues shifts the equilibrium toward the inactive complex, and oxygen is released.[21]

Evolutionary Modifications

Hemoglobin is undoubtedly the most complete example of how structure affects physiology and how both structure and physiology have been modified in evolution. Physicians have identified many varieties of human hemoglobin that are partially defective in function and cause disease. The first disease ever characterized at the level of atomic structure was sickle cell anemia, which is caused by a single amino acid change. In this section we look not at the pathologies of hemoglobin but at the evolutionary modifications where the equilibrium between the active and inactive forms of hemoglobin has been altered.

In evolution, hemoglobin has been subject to many modifications that allow animals to adapt to new physiological conditions. The special adaptations of mammals have generated three solutions to the problems of placental development, in which the fetus, located far from the mother's lungs, competes with maternal tissues to get adequate oxygen. Usually in mammals, the fetal red blood cells produce less diphosphoglycerate than the maternal red blood cells. The hemoglobin of the fetal red blood cell therefore is left in its active state, allowing it to load oxygen at the expense of the surrounding maternal tissues. This simple physiologic adaptation uses the same machinery that is available for adult physiology.

Primates have evolved a second hemoglobin, a genetic variant of the first, for use in the fetus. The principal alpha and beta hemoglobin chains arose about 450 million years ago in fish, when the ancestral globin gene duplicated and diverged in sequence. Then the gene for the beta chain duplicated again and further diverged 150 million years ago during the Jurassic, producing the fetal variant in the line of early mammals that would later lead to primates. This fetal hemoglobin has a slightly different amino acid sequence. The human fetal hemoglobin has about the same oxygen affinity as adult hemoglobin, but because of a few changes in its amino acid sequence, it binds diphosphoglycerate poorly and tends to remain in the active state. Since diphosphoglycerate in the maternal circulation drives the equilibrium of maternal hemoglobin to the inactive state, the human fetal hemoglobin can rob oxygen from the maternal circulation, because it is insensitive to diphosphoglycerate.[22]

In cattle, sheep, and goats, fetal hemoglobin of yet a different kind was generated from a duplicated beta gene about 50 million years ago. In these animals, the adult hemoglobin is affected not by diphosphoglycerate but by phosphate in the red blood cell, in the way that favors oxygen unloading. The fetal hemoglobin has an intrinsically higher affinity for oxygen, in part because the inactive state is not stabilized by phosphate in the fetal red blood cell. It is therefore shifted to the active state, again allowing the fetus to win out over the maternal hemoglobin of the ruminant. It is not surprising that there are many ways to shift the hemoglobin equilibrium, all with the same result, once the two-state equilibrium evolved in the first place.

One of the most striking heritable adaptations in hemoglobin is the single amino acid change in the hemoglobin of the bar-headed

goose that flies over the Himalayan mountains at 30,000 feet (9,200 m), an altitude with only 29 percent of the oxygen at sea level. A single amino acid change in the goose's hemoglobin, as compared to that of its lowland relative, shifts the population toward the active state in the absence of oxygen. Hence the hemoglobin loads better at low oxygen tension. If that same mutation occurs in human hemoglobin, it too shifts to the active state. In some invertebrates there is, as in vertebrate hemoglobin, a sequential expression of different hemoglobins with different affinities for oxygen at different stages of the life cycle. It is not known how these affinities are generated, but it might well be done by shifting a two-state equilibrium via amino acid substitutions that are genetically based.[23]

Physiological and Evolutionary Adaptation

It is obvious from these examples that animals can use two strategies, one genetic and one physiological (somatic), to modify the oxygen affinity of hemoglobin in response to hypoxia, including the form of hypoxia normally encountered by the fetus. Both mechanisms rely on shifting a preexisting two-state equilibrium one way or the other between the inherently active and inactive states, that is, between states that excel either at loading oxygen or at unloading oxygen. If we think about the great inventions in evolution, the concerted two-state transition was assuredly one of them. It is very ancient. It is very conserved. And it is very modifiable. Most important, this modifiability is selected in each generation of the organism for the physiological adaptations that it allows.

To add to the examples above, consider the hemoglobin of the adult llama, which lives at high altitudes in the Andes. Mutational changes have caused it to have a reduced affinity for small molecules containing phosphate, much like primate fetal hemoglobin. Yet a strain of domesticated chicken of the same altitude and region has the same hemoglobin as lowland strains but its level of organic phosphates in the blood cell is lower, owing to mutation. These adaptations can be accomplished in numerous ways, each a small structural step that yields major benefits to the organism.[24]

The evolutionary modifications of hemoglobin are reminiscent of sex determination in showing how easily genetic control and physiological control can substitute for each other. In the two-state system of hemoglobin, a single mutational change can replace external regulation, as in the case of the high-flying goose, where a single amino acid change creates a new physiology. The hemoglobin molecule is a poised system; either environmental inputs or mutation can trigger a change of oxygen loading or unloading. Evolution of new physiologies, such as acclimation to high altitudes and fetal hemoglobin, is only a step or two away.

As Baldwin and Schmalhausen predicted, the particular somatic adaptability was probably there first in the cases discussed here, before a mutation arose to fix and perhaps extend one of the alternative states of that adaptability. It is difficult to prove the Baldwin scenario, though, for any of them. We can imagine that the evolutionary change in hemoglobin in the high-flying goose might have come about when normal geese with their normal two-state hemoglobin searched for food or migration routes, going as high in altitude as they could in the Himalayas with their normal oxygen-carrying adaptability. Those conditions would result in a strong directional selection for single amino acid changes to increase the oxygen avidity of hemoglobin. The intrinsically active (high-affinity) state was stabilized, making use of what was already there.

By this means, the somatic adaptation became an evolutionary adaptation. A modest mutational investment had a big phenotypic payoff: success in migrating over Mount Everest. With time, the development and functional output of the phenotypic change may undergo additional variation and selection such that its development and function are narrowed to an optimal range—a further stabilizing selection, according to Schmalhausen.

The molecular understanding of hemoglobin has allowed us to go far beyond Schmalhausen and Baldwin. What has been added here is that evolution can build on physiology by acting on highly poised, switch-like systems, which themselves are highly constrained and conserved. The phenotype (oxygen transport) is divided into (1) a conserved adaptive system of some complexity, like the hemoglobin protein and its dynamic equilibrium, and (2) simple signals like oxygen and diphosphoglycerate. This configuration facilitates genetic replace-

ment of the simple signals with simple mutations of protein structure, and in this way small genetic variation can have major effects on phenotypic variation.

Somatic Adaptation and Evolution

The lesson from Baldwin, Schmalhausen, and Waddington is that the organism has a great deal of latent novelty within its own somatic adaptability. As West-Eberhard has extended the lesson, all phenotypic novelties are reorganizations of preexisting phenotypes. In effect, the organism can express many alternative phenotypes—phenotypes that are stable like the different insect castes and the male-female alternatives of sex determination, or phenotypes that are readily reversible like the alternative oxygen-loading and oxygen-unloading forms of hemoglobin. Because they have already been tested in evolution, these phenotypes are necessarily viable and adaptive to the ambient conditions. They are a special set of phenotypes and anything but random.

Some of these somatic adaptations are simple whereas others, like the various forms of developmental plasticity, incorporate an entire suite of anatomies, physiologies, and behaviors. These alternative paths of development can be easily stabilized and modified, in part because they normally require signals to proceed from one state to another. Omitting or inhibiting signals is readily achieved by simple mutations, an example of gene-environment interchangeability, which has been documented on the morphological level in both natural and laboratory studies.[25]

To counter the argument that stabilization of the somatic adaptations is an important means of facilitating evolutionary adaptation, a skeptic might argue that not much new has been achieved. The modifications might be characterized as modest, sometimes just a simplification of the complexity that was built up in evolution. To choose one of the alternative phenotypes of the honeybee for further evolutionary elaboration is not to invent that phenotype in the first place. The requirement that abundant novelty must already exist within the organism could be a real limitation on these ideas, or at worst a prescription that evolution always proceeds from the complex to the simple.

Yet the examples described here are not simplifications. Even in the mechanistically obvious case of hemoglobin, the development of a placental physiology dependent on a new form of hemoglobin physiology is not an elaboration of what is already there. It is an exploitation of the capacity of the hemoglobin molecule to modify its physiological range by mutation or genetic reassortment from existing genetic variation in the population. The capacity of hemoglobin to be modifiable was not selected for ease of future evolutionary genetic modification, but for reversible modification by environmental conditions, which has value in each generation.

Although much of biology may seem ad hoc—polyphenism in bees, sex determination in reptiles, hemoglobin physiology—underneath these processes are some very general and ancient mechanisms. As we shall see, hemoglobin is but one of many examples of highly poised proteins that, unlike hemoglobin, function in all the cells of our body. Adaptability is a key characteristic of many of the conserved core processes of eukaryotes. Polyphenism and environmental sex determination make use of highly modifiable transcriptional mechanisms (another set of core processes), which also play a large role in embryonic development.

It may seem counterintuitive that mechanisms such as oxygen regulation, which function to maintain the existing phenotype by buffering the effects of variation in the environment, should simultaneously serve as vehicles for creating variation in evolution. This pseudoparadox of stability versus change stands juxtaposed to another of conservation and diversity. How do highly conserved processes like oxygen transport in hemoglobin or determination of sex in mammals lower the barrier for the generation of diversity?

The answer to both apparent paradoxes is quite obvious when one examines them at the molecular level. The organism is not robust because it has been built in such a rigid manner that it does not buckle under stress. Its robustness stems from a physiology that is adaptive. It stays the same, not because it cannot change but because it compensates for change around it. The secret of the stability of the phenotype is dynamic restoration. Mutations or genetic reassortments that target these dynamic restorative systems can reset their optima and generate a class of significant phenotypes with reduced lethality. Evolution can achieve new forms of somatic adaptation so readily because the system, at all levels, is built to vary.

FOUR

Weak Regulatory Linkage

Let us now delve directly into the conserved core processes that are responsible for generating most of the anatomy, physiology, and behavior of the organism. These are the processes that evolved between three billion and a half billion years ago (Chapter 2). They include metabolism, gene expression, and signaling between cells.

In deepening the inquiry, we introduce the fact that all the conserved core processes possess adaptability, which they use in response to varying conditions inside the organism rather than to identifiable external conditions, although they can do that as well. We ask the same question as before: How can these core processes facilitate the generation of novel variation in evolution?

The core processes reveal conservation and economy. On the conservation side is evidence that the genes encoding the RNAs and proteins of these processes are highly conserved across diverse animals, from jellyfish to humans. We have seen that the core processes were established in several great waves of innovation, and since then they have remained basically unchanged. On the economy side is recognition of just how few genes a complex organism has to work with—only 22,500 in humans, about one and a half times those of a fruit fly.

These two facts, conservation and economy, suggest that complexity must arise through the multiple use of a relatively few conserved elements. Complexity arises when different parts of the adaptive range are selected. It also arises when different combinations of conserved elements are chosen. Conservation can be compatible with economy, as long as the elements of the core processes have properties that allow them to make diverse combinations with differing consequences. For the organism to generate different cell types and different cell behaviors, it must produce and respond to diverse signals, retrieve diverse information from the genome, and generate diverse combinations of cell behaviors. The ability to process all this information with a limited set of components underlies the somatic adaptability of the core processes.

In evolution, these preexisting combinations of cell behaviors and expressed genes must be altered to give new combinations of behaviors and genes. Some kinds of somatic adaptability arise by modifying the way the conserved core processes are linked to one another. By linkage or, better, by regulatory linkage, we usually mean how information is passed from one component to another. Since all information transfer occurs on the molecular level, we mean specifically how one molecule passes information to another. Signals may come from outside the cell or the organism. They must be passed through a chain of command until a response is made. That response could then affect the environment or the organism.

Eating a candy bar produces an input of sugar. After the body takes up sugar and senses the level by a complex pathway, the pancreas responds appropriately by releasing insulin. Increased insulin in the blood elicits diverse responses in the tissues: making fat in fat cells, making glycogen in the liver, and taking up sugar in the muscles. Relatively small molecules, such as sugar or insulin, cannot directly cause such disparate and complex effects. Sugar has to act indirectly to elicit insulin secretion, and insulin has to act indirectly to elicit the diverse cellular responses.

To mediate these effects, linkages must exist at nearly every step between sugar and insulin and between insulin and tissue responses. How the linkages are crafted is very important, for they tell us how new linkages can be generated in evolution. Throughout biology, it turns out, individual core processes are constructed so that new linkages can easily be forged and broken. As in hemoglobin, new physiologies and behaviors can arise that require little genetic change to the existing complexity of the organism.

In this chapter we use the term *weak linkage*, first coined by Michael Conrad in 1990, to mean an indirect, undemanding, low-information kind of regulatory connection, one that can be easily broken or redirected for other purposes. It may also be physically weak, but the main emphasis is on its minimal and readily changeable nature. A simple analogy might be electrical plugs and sockets, which because of their standardized design can accommodate many different connections. Here we consider linkages, not as arrows on some organizational chart, but on a molecular level where their special properties of facile reorganization can be appreciated.[1]

The pivotal insights into the molecular character of linkages in these biological information circuits came with research on gene function in bacteria in the 1950s and 1960s. The experiments of Jacques Monod and François Jacob are legendary in the history of molecular biology, but their study of gene function also had strong implications for evolution. They explained how biology simultaneously achieves versatile usage and physiological robustness in the regulation of metabolism by small molecules.

Jacob and Monod studied bacteria, whose circuits are not as complex as those in multicellular organisms. Nevertheless, bacteria use weak regulatory linkage in their relatively simple circuits. Eukaryotic cells have added more complexity and versatility to these linkages, but the principles established in bacteria are fundamental to animals as well. The studies of bacterial metabolism were representative of much complex biology, because they concerned adaptation—how the phenotype of an organism can be controlled by small molecules. Also, these studies were the first to deal with the process of gene regulation.

Understanding embryonic development is central for explaining phenotypic novelty in animals. It is in the embryo that much of the phenotype is established with all its anatomical and physiological complexity. It is in the embryo that we might look for those "adaptive cell behaviors" that Sewall Wright asserted underlie the generation of phenotypic variety. The work on bacteria became surprisingly relevant because, as in metabolism, many complicated processes of embryonic development also depend on relatively simple signals mediated by rather small molecules. For bacteria, which are single-celled organisms, the small molecule signals are merely ingredients of the environment. But in multicellular organisms, many of the relevant small molecules are signals synthesized by cells of the organism itself and passed to other cells. Though simple in composition, these signals may regulate the most complicated developmental circuits, such as those that give rise to the entire nervous system. The implications for evolution are powerful, for if complex development is elicited by simple signals, then changes in complex development may be achieved by changing the amount or the location of these simple signals, rather than by changing a highly integrated and complex process.

Although studying the properties of these regulatory pathways is informative, we ultimately want to understand at an even more chemical level what makes the linkages weak. (Remember, by weak we mean minimal and easily changeable.) The features that allow for invention and reorganization can best be appreciated on the atomic level.

Here again, Jacob and Monod offered the first and arguably the most general ideas, and they provided regulatory weak linkage its first and most enduring molecular mechanism. It is a mechanism we have already seen in the special case of hemoglobin. Weak linkage, we shall see, is not a core process; but the capacity for such linkage *is* a core property of the processes. It underlies the mechanism of many core processes, such as transcription and signaling between cells. Although the exact chemistry may vary, all core processes have the capacity to be weakly linked to other processes and conditions.

Control of Gene Function

Most of biochemistry until the mid twentieth century concerned metabolism—the breakdown of foodstuffs to extract energy, and the synthesis of the components of the body. Little attention was paid to how that metabolism was regulated in response to internal needs or external opportunities. Regulatory biology came of age with DNA and an understanding of protein synthesis. The early forays of molecular biology into regulation gave the first molecular picture of how time, place, amount, and circumstances controlled what the cell did or what the cell produced. In other words, the questions focused on how the organism could change its phenotype without changing its genome. With the early knowledge of regulation came questions of how that regulation could change in evolution. Thus molecular physiology was born and gave impetus to an understanding of facilitated variation.

We owe a lot of that early molecular understanding to the French molecular biologist Jacques Monod (1910–1976). More than any other scientist, he united genetic and physiological mechanisms at the most basic molecular level. A passionate and sometimes overbearing man, Monod was at the epicenter of a remarkable group of French scientists at the Pasteur Institute after World War II. Monod came to the regulation problem through an interest in metabolism, which was the most advanced biochemical subject of the time. The exact molecular structure of most metabolites was known by midcentury, as well as the enzyme-catalyzed steps for generating and using them.

Growing the bacterium *Escherichia coli* on different sugars, Monod encountered a real paradox, a very fortunate one. The growth rate of *E. coli* on a mixture of two sugars was not the sum of the rates on either one alone. The bacterial culture would first grow for a while using one of the sugars, then pause, and then start growing again using the other sugar, as shown in Figure 17.

To grow on a sugar, the bacterium had to have a particular enzyme to degrade it. Monod found that the bacterium did not have that enzyme initially. It first produced one kind of enzyme to metabolize one of the sugars, and then produced a different kind to metabolize the other. This obscure phenomenon, unknown to Monod when he started his experiments, had been known to others since 1901. "Enzyme adaptation" referred to the experimental fact that the parsimonious bacterium adjusted its metabolism according to the food supply.

The explanation for Monod's paradox of why the bacterium consumed the sugars sequentially rather than simultaneously was that one

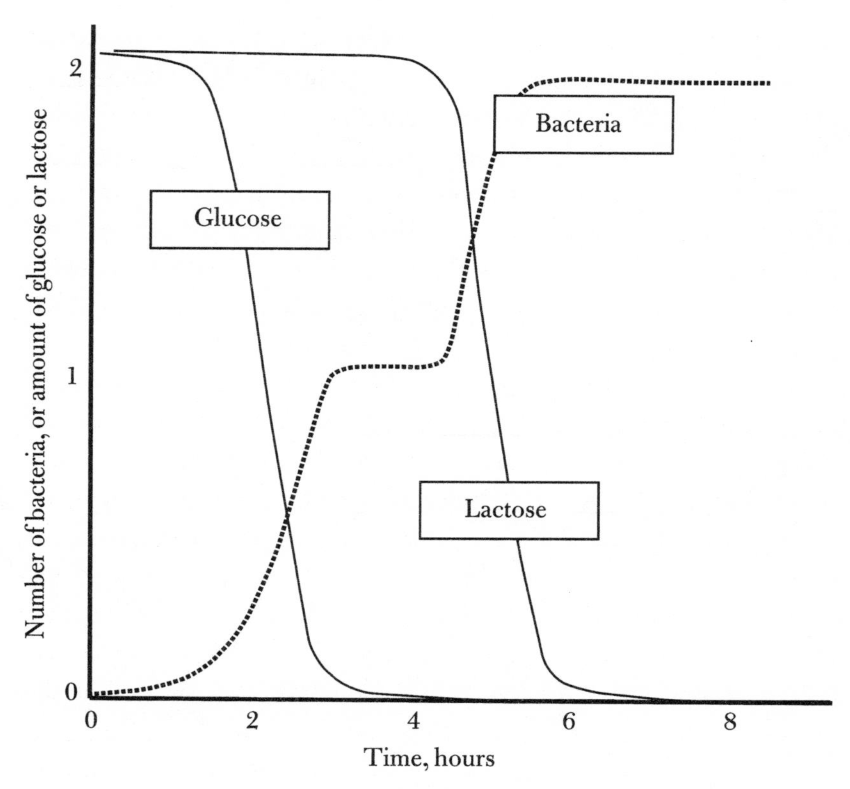

Figure 17 The Monod experiment. Bacteria grow in nutrient liquid containing two sugars, glucose and lactose. They preferentially use glucose first. When it is gone, they stop growing for an hour and adapt to using lactose by making a new enzyme, beta-galactosidase. Then they grow again until the lactose is gone. The analysis of this "enzyme adaptation" by Jacques Monod led to our modern understanding of the control of gene expression in all organisms.

sugar preferentially stimulated the bacterium to produce an enzyme to degrade it, while also inhibiting the bacterium from producing the enzyme to metabolize the other sugar. Here was an exquisite form of physiological adaptation, the organism fastidiously changing the proteins it produced (part of its phenotype) in response to changing environmental conditions.[2]

For further study, Jacques Monod chose the milk sugar lactose and the enzyme that metabolized it, lactase (now used for removing lactose from dairy products for lactose-intolerant people), also known as beta-galactosidase. *E. coli*, in the gut of an infant, would encounter lactose after the infant nursed and would soon be ready to metabolize it at a furious rate. In the intervals between feeding, it would turn off the production of the enzyme and metabolize other sugars. Monod found that beta-galactosidase was not stored in an inactive form but each time was synthesized anew from available amino acids. Thus, enzyme adaptation was not a process of *activation* of an enzyme but a process of *synthesis* of the enzyme. In the end, the regulation was on the synthesis of the RNA transcript from the DNA. The enzyme itself was not adapting but the bacterium was, by producing more or less of the specific enzyme. The phenomenon was renamed enzyme induction.

Although this might seem to be merely another example of physiological adaptation (like oxygen uptake responding to oxygen supply), it was special because it involved the synthesis of a protein that would ultimately be connected to the genes that encode that protein. At the time, it was not known how an organism could selectively make one kind of protein of the many that were encoded in its DNA. It was possible that every protein had its own induction mechanism, but it was much more likely that most proteins were made continuously. Inducible proteins like beta-galactosidase would be exceptional by being *not* synthesized at specific times.

Monod theorized that the synthesis of each inducible protein would be inhibited by a specific inhibitory protein, the *repressor*, which the cell made continuously. Lactose, when present, would bind to the repressor and remove it from the DNA, in this way releasing the inhibition. In this view there were no real inducers per se, only repressors. Activation was then achieved by the antagonism of repression, or as Monod called it, derepression. He was adamant that derepression is the only possibility; however, his colleague François Jacob speculated correctly that true activators might exist that could bind to DNA. They would activate specific genes to produce the corresponding RNA, which would yield the protein.

Biology's solution to the selective use of information from the genome was elegantly simple. The bacterial cell could adapt to changes in the environment by repressing and derepressing genes in a simple stimulus-response circuit. Jacob and Monod's experiments proved the model to satisfaction in bacteria. For example, they isolated cells defective in the gene encoding the repressor protein. These cells made beta-galactosidase all the time, regardless of whether lactose was in the environment.

This simple physiological circuit was proposed as a general solution for regulating the expression of different genes in different cells. In the spirit of the new generalizations of the day, such as the universal structure of DNA, the universal genetic code, and the universal steps of metabolism, there was no doubt in the minds of Jacob and Monod that they had discovered another core process, which would have unlimited applicability—the process of "regulation of gene expression."

Monod worried about one part of the model. Physiological responses are usually quantitative, showing smooth and continuous variation, not on-off extremes. This trait was true of lactose metabolism. Over a wide range, the more lactose in the medium, the more enzyme the bacterium made. What the two men had described instead was an on-off switch, a poor model for physiology. Jacob was able to resolve this theoretical difficulty: "The insight had come to me while I watched one of my sons playing with a small electric train. Although he did not have a rheostat, he could make the train travel at different but constant speeds just by turning the switch on and off more or less rapidly."

Physiological adaptation in bacterial enzyme synthesis was now largely solved in a wonderfully simple manner, a genetic switch, turning the gene off by binding a repressor to it and turning it on by removing the repressor with an inducer. The fraction of time the switch was on would determine the rate of synthesis. The details were soon filled in: the repressor protein of beta-galactosidase synthesis binds to a specific short sequence on the DNA, located next to the start of the beta-galactosidase gene sequence, where it occludes the binding of the machinery necessary to transcribe the RNA (Figure 18).[3]

There are two linkages in this system. First, the repressor binds to the DNA at a specific location and blocks RNA polymerase from

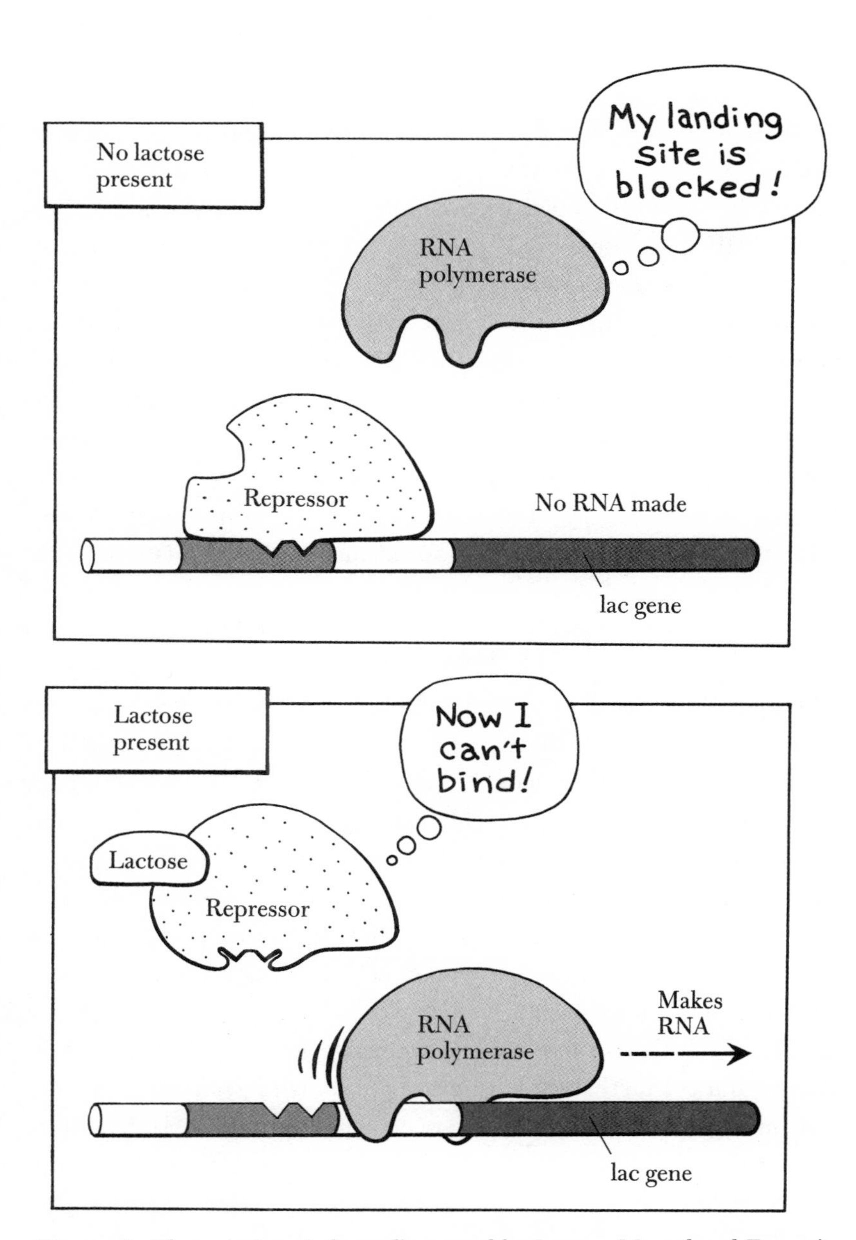

Figure 18 The genetic switch, as discovered by Jacques Monod and François Jacob. A specific repressor protein acts as the switch. When it binds to a DNA site near the gene encoding beta-galactosidase, the RNA polymerase protein cannot bind nor can it synthesize RNA from the gene. The gene is turned off. When lactose is present, it binds to the repressor and keeps it from the DNA site. The gene turns on.

synthesizing RNA from that gene only. Second, the lactose binds to the repressor and causes the repressor to change so that it no longer binds to the DNA. Though several molecules like lactose regulate their own metabolism, none binds to DNA directly; they all act indirectly through a repressor protein.

In such an indirect system, the response can be easily modified and generalized. If the lactose repressor were to bind next to some other gene, that gene should also respond to lactose. In fact that is the case for a specific transport protein for lactose, whose synthesis is then coordinately regulated with beta-galactosidase. The indirectness of the linkage between the signal, the lactose, and the response (RNA synthesis), offers many opportunities for changing gene regulation.

Weakening the Linkage

Although this was a tidy model for adaptive sugar metabolism in bacteria, it did not yet qualify as a conserved core process central to all multicellular development. Yet the model soon served as a framework for understanding the regulation of genes in all of biology, even if the exact mechanisms differed. The problem of regulating gene expression is particularly acute in multicellular organisms. Humans, for example, have more than three hundred recognizable cell types and consist of about one hundred trillion cells in complex arrangements. By expressing a unique subset of genes, each cell type makes a unique profile of the kinds of proteins responsible for its unique activities.

Although it is true that we have more genes than *E. coli*, we have only six times more than that simple single-celled organism. To an extent that far exceeds that for bacterial genes, human genes are read out in different places, at different times, and in different circumstances, presumably in many different combinations.

In the end, several major refinements had to be added to the bacterial model to explain gene regulation in the much more complicated eukaryotic cells, including humans. These refinements make it easier to generate more complexity and also easier to change that complexity in different circumstances or locations in the animal.

The first refinement added positively acting processes to the powerful repressive mechanisms of Jacob and Monod. Together, the positively and negatively acting proteins are called transcriptional regulators, or transcription factors. A second refinement was the linkage of several transcriptional regulators and genes into complex circuits, including circuits in which certain regulators control the expression of genes encoding other regulators. These circuits can have logical and operational features like those in computers. Biology developed in the direction of linking together simple circuits, rather than making the individual circuits more complex. The logical structure of these circuits is only now being worked out, but they bear strong resemblance to logic circuits in engineering. A third addition was the different organization of genes in eukaryotes that allowed for forms of regulation not found in bacteria. Finally, the bacterial geneticists of the 1950s and 1960s were oblivious to the extensive control of protein levels and activities that occur, not at the level of transcription in the nucleus, but in the cytoplasm of eukaryotic cells. They are controlled through regulated protein modification leading to activation or degradation of the protein and RNA.[4]

With a more complete tool kit for regulating protein levels and protein activity, the eukaryotic cell, sometime between two billion and one billion years ago, had achieved processes of sufficient power to regulate large combinations of genes through extremely complex circuits. Except for elaboration and diversification of the circuits, very little of the process has changed qualitatively since. These regulatory mechanisms apparently possessed enough power and versatility to facilitate all of the evolution that has occurred since the time of their invention, while they were themselves conserved. Therefore, eukaryotic gene regulation is perhaps the most powerful conserved core process, responsible for much of the phenotypic variation on which selection acts.

In eukaryotes, various regulatory proteins commonly bind to specific regions of DNA sequence near the gene to be controlled, and directly activate transcription in response to extracellular signals, as opposed to removing a block to transcription, as in the Jacob-Monod mechanism of derepression. For example, thyroid hormone regulates many genes in humans, such as those related to growth. It is made in the thyroid gland and transported through the blood to the cells of the body. It then enters the nucleus of a cell and binds to a protein that, like a true repressor, is associated with the DNA and blocks transcription (with the help of still other proteins). Binding the hormone does not remove that thyroid receptor from the DNA but instead turns it into an activator, which now remains on the DNA and interacts positively with the machinery for making RNA.

The typical mammalian gene is regulated by dozens of such factors binding in the vicinity of the gene at particular DNA sequences they recognize. Each carries a message from some different signaling system, specifying time, place, amount, cell type, or other information. These signals are not on-off switches but partial-on and partial-off switches. The multiple factors result in a certain level of RNA synthesis from a specific gene.[5]

Other Forms of Remote Control

The control of gene expression in eukaryotic cells and in particular in multicellular organisms has weakened the linkage between signals and transcriptional responses in two ways. Both are part of the process of enlarging the control regions of DNA. The first relaxes the geometric requirements for the interaction of the repressor with the transcriptional machinery. In bacteria, the repressor has to bind precisely to a particular site on the DNA to be in position to physically block the binding of the enzyme that makes the RNA transcript, as shown in Figure 18. Movement of the binding site even a short distance undermines this regulation. Transcriptional control in eukaryotes, as mentioned above, is the epitome of weak linage. Protein factors do not have to be positioned on the DNA with precision but need only bind in the vicinity of the gene; they can be either in front of or behind the start site for RNA synthesis. It is easy to evolve a new regulatory feature of a gene: all it takes is to bind a transcription factor anywhere near the gene.

The regulatory regions accompanying the genes of multicellular organisms can be enormous in sequence length, encompassing a thousand times more DNA than in the bacterial examples. The regulation of eukaryotic transcription is far less precisely organized than the systems in prokaryotes. The placement of regulatory factors seems haphazard. This tolerance of "sloppy" placement is, however, a feature that lowers the barrier for the incorporation of new sequences and for the generation of phenotypic variation on the level of gene expression.

The second feature that weakens the linkage is that some DNA-binding proteins that bind to sequences near the gene do not have to "touch" the transcriptional machinery at all. Instead of interacting directly with the machinery, they bring enzymes to the vicinity of the gene that in turn alter the structure of the proteins that surround the gene. The activity of the gene is eventually based on a "consensus" reached by a large number of contending inputs, some of which activate and some of which inhibit the gene, but none of which has to make direct chemical contact with the machinery for RNA synthesis.

The Cell Response to Signals

The regulation of gene expression is one of the core processes most critical for generating phenotypic variation. The ease of remodeling the linkages that regulate genes is directly related to the ease of generating novel patterns of gene expression in evolution. For multicellular organisms, the major arena for anatomical variation is the processes of embryonic development. Spatially and temporally regulated patterns of gene expression drive these processes. It is now possible to connect weak linkage more directly to the expression of individual genes in order to answer two questions: How is the anatomy of multicellular organisms generated in development? How might development change in evolution?

Jacob and Monod hoped that the simple explanation of enzyme induction in bacteria could serve as a model for the complex process of selectively reading out genes from the genome during embryonic development. There the inducers would not be sugars in the environment but signals generated by other cells in the embryo. The two men showed theoretically that complex circuits could be built with simple switches connected in various ways. Though the fields of embryology and genetics had diverged sharply at the turn of the twentieth century, interest finally converged on two questions concerning gene expression: How does a single cell, the fertilized egg, with a single genome, give rise to many different cells in the embryo and adult? How does each cell type produce a unique set of proteins for its unique function? An obvious follow-up question is, How readily can patterns of gene expression change in evolution?[6]

Embryologists thought the questions could only be addressed in embryos. Geneticists and biochemists were content to infer the explanation from the study of simpler, more tractable systems. In the end, both proved to be correct. As early as 1934, T. H. Morgan sketched a partial answer: "The initial differences in the protoplasmic regions [of the embryo] may be supposed to affect the activity of genes. The genes will then in turn affect the protoplasm, which will start a new series of reciprocal reactions. In this way, we can picture to ourselves the gradual elaboration and differentiation of the various regions of the embryo."[7]

This model sounds very similar to a sequential enzyme induction, where the inducers would be materials in cells or made by cells, and the induced enzymes would be all the proteins specific to the differentiated cell state. Was the simple model of enzyme induction in bacteria indeed the answer to the larger question of how an animal develops? If so, concepts like weak linkage would bear directly on the feasibility of phenotypic variation, for it is the phenotype that is being constructed in the development of the embryo.

The golden age of embryology occurred in the period 1900–1930, and its greatest accomplishment came in 1924 with Hans Spemann and Hilde Mangold's discovery of "embryonic induction." In the experiment illustrated in Figure 19, Spemann and Mangold isolated a small piece of tissue from a newt embryo at an early stage well before cells were differentiated. From that region the embryo would later develop its trunk and back. When they transplanted this small region into a recipient of the same age, at a site that would have normally developed

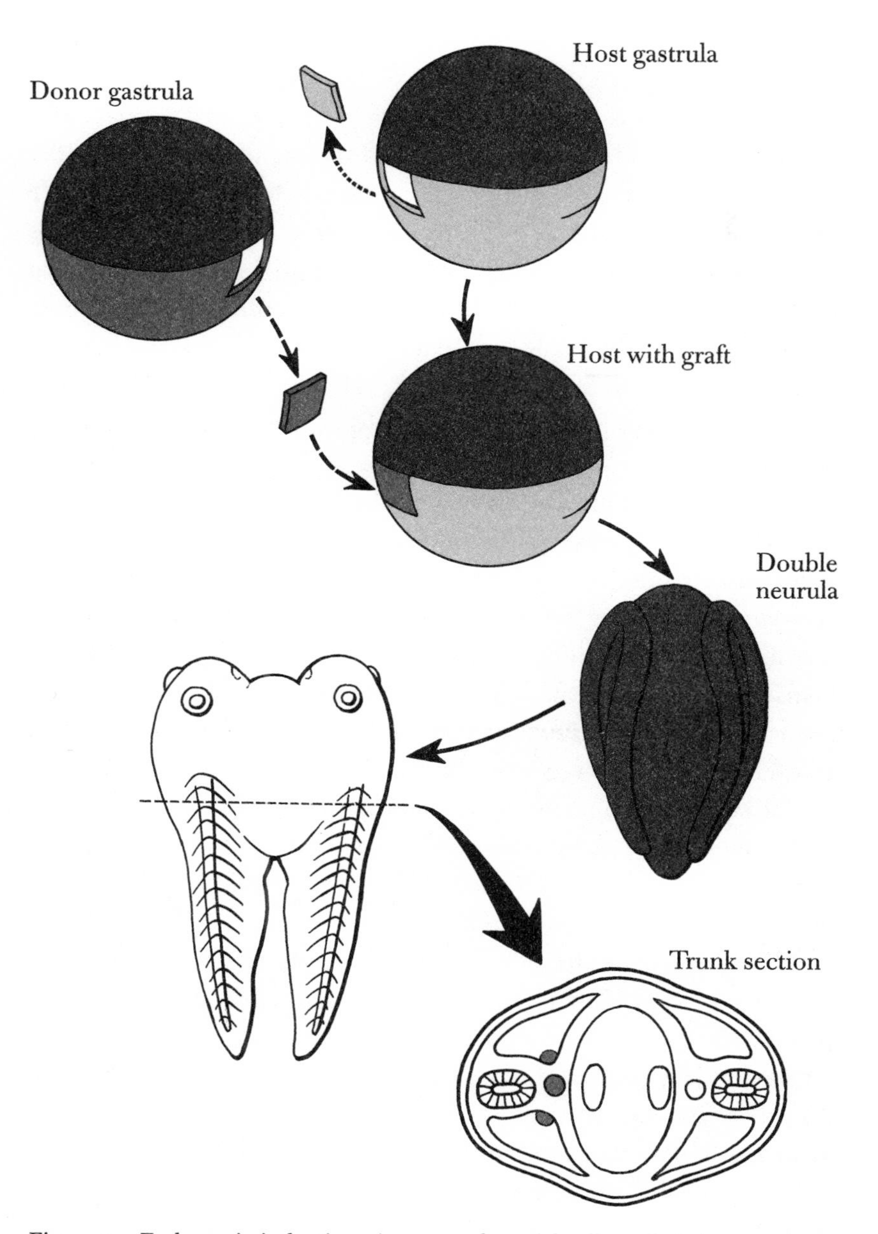

Figure 19 Embryonic induction. A group of special cells, called the organizer, releases inducer proteins that trigger the development of the nervous system and the musculature in its vicinity. The organizer activity is assayed in the grafting experiment of Hans Spemann and Hilde Mangold shown here. The gastrula-stage frog or salamander embryos are oriented with the organizer cells on the right, recognized by the horizontal streak on the embryo. After the operation the host embryo, with two organizers, develops a neurula-stage embryo with the beginnings of two nervous systems, and then becomes a conjoined twin. In the cross-section of the twin, the transplanted organizer has contributed only a few cells to the body axis on the left, the dark spots. The remainder of that axis has been induced by the grafted organizer.

into the embryo's belly, the transplanted piece "induced" the nearby cells to forgo belly development and instead to form virtually a whole new embryo at the site. The host embryo with its graft developed as conjoined twins with two complete heads (the original and the induced one), two spinal cords, and two blocks of skeletal muscle. Since little of the grafted tissue was incorporated into the new embryonic structures, the researchers concluded that the grafted tissue had induced the new embryonic parts from the host tissue, which on its own would have made belly.[8]

Only a small region of the embryo had the power of induction. It was called the organizer of the animal's body axis, to acknowledge its indispensable role. By the 1960s, the word *induction* had gained two very specific meanings. In embryology, it meant "the determination of the development or differentiation of an embryonic region into a particular morphogenetic pattern by the influence or activity of another embryonic region." In biochemistry, it meant "an increase in the rate at which an enzyme is synthesized by a cell (especially in a microorganism), or the initiation of its synthesis, as a result of the exposure of the cell to some specific substance (the inducer)."[9] Might these very different definitions of the same word suggest that they are not merely homonyms?

The search for the elusive embryonic inducer consumed mid-twentieth-century embryology. The transplanted tissue seemed to impart essential and specific information to transform belly precursors into a normally patterned brain and nervous system. Originally it was thought that the graft had to be living and intact to exert its effect. With the finding that minced, heated, and ground-up tissue still had inducing activity, the search quickly shifted to chemical signals. Embryologists became the present-day alchemists trying exotic crude preparations of foreign tissues, such as guinea-pig bone marrow or fish swim bladder. Amazingly, many of these tissues contained inducing substances. Even simple toxic chemicals such as ammonia and acid worked.

The search for inducers collapsed in confusion with the growing awareness that the inducers might not need to provide specific chemical information. When experimental embryology was reborn forty years later in the mid-1980s, the experimental approaches came from rather different directions: from genetic studies of developing fruit flies, from the new molecular biology, and from the successful isolation of the first signaling proteins ("growth factors") from mammals. By 1995, the organizer inductive activity was known to include a small set of secreted proteins. These had specific functions, which unexpectedly resembled formally the derepression phenomenon of bacteria.

Two extreme views of information transfer have always existed in biology, the permissive and the instructive. The distinction comes up whenever there is a stimulus and a response, or more generally a cause and an effect. For a particular response or effect, how much information is provided by the stimulus, the seeming cause, to get an effect? Watering a seed provides a stimulus, but it is a permissive input, since no one would assume that the water falling on the seed instructs the seed how to germinate into a plant. In contrast, when Gilbert and Sullivan collaborated on *The Mikado*, we assume each contributed important talents to the outcome. Gilbert did not just serve tea to Sullivan, who then wrote both the lyrics and the score. If Gilbert's lyrics provoked Sullivan's music, the process was clearly instructive; Gilbert provided crucial information, but of course not *all* the information. Obviously, permissiveness and instructiveness are a matter of degree, a measure of exactly what each agent contributes to the response.

One of the crucial implications of the Jacob and Monod model was that it explained how a complex response could result from a simple and permissive signal. Lactose is a simple signal for a very complicated response: production of the enzyme beta-galactosidase, composed of over four thousand amino acids strung together in a specific sequence. Does lactose inform the cell how to make the enzyme? No, the cell's synthetic system, from the gene on up, is already complete and poised to make the enzyme. It is held back by the repressor protein. The lactose-inducing signal merely releases the block imposed by the repressor (itself a complicated structure). It is like pouring water on the seed. Once the repressor is removed from the DNA, the transcriptional machinery does what it has been designed to do. Thus, beta-galactosidase induction by lactose is a permissive interaction.

Although permissive signaling works for regulating beta-galactosidase, it seemed unlikely to many biologists that the signals of embryonic induction could act permissively. The outcome of embryonic induction is, after all, extremely complicated; it results in creation of virtually the entire organized embryo, with hundreds of cell types and many organs, including the entire nervous system, all in the right places. The name "organizer" that Spemann gave to the source of the inducer implied instruction, maybe even micromanagement. To everyone's surprise, embryonic induction turned out to be a permissive process; the organizer provides a signal of little complexity.

In the newt, all regions of the embryo are initially capable of developing into almost anything: the nervous system, vertebral column, or muscles of the head, trunk, or tail. This potentiality is *globally repressed* by a signaling protein and other factors, which all cells secrete and communally receive. They constitute the formal equivalent of the repressor described previously, except that the signaling protein is not directly a repressor of genes; it acts at the cell surface through a pathway (involving several relay intermediates) that leads eventually to gene regulation, both repression and activation. This repression is different from beta-galactosidase in that it is connected through a hierarchy of signals to several circuits that generate additional secreted signals, which in turn stimulate other more complicated responses.

What, then, is the embryonic inducer? It is simply a collection of secreted proteins that locally bind up and antagonize the ubiquitous repressor signal. In the vicinity of these antagonists, the embryonic cells are released from their self-imposed repression, and develop the nervous system, vertebrae, and musculature of the back side of the body. Normally, the antagonists from the organizer do not reach the opposite side of the embryo, and that is where the still-repressed cells develop the belly. When the inducing tissue is transplanted to a second embryo at a location that normally develops the belly, a second realm of antagonism is established and the second embryo develops dual nervous systems, vertebrae, and muscles. The inducers merely release the innate but self-inhibited capacity to develop these structures.

In retrospect, the disconcerting finding that simple chemicals, foreign to the embryo, can evoke the entire nervous system must reflect how easy it is to antagonize the ubiquitous repressor signal and how inherently competent the cells are to pursue complex development.[10]

What might have been the alternative to permissive signaling? In the view of many embryologists before the mechanism of induction was elucidated, the signaling molecules in embryonic development should provide information critical for the process to proceed. DNA or RNA might have been passed from signaling to responding cells, "instructing" them on the next steps of development, providing information utterly outside their ken. Alternatively, signals might have been released as a complex spatially arranged code. Or enzymes might have entered the cell and there carried out new transformations. These alternatives have never been found to occur. Signaling molecules exchanged between tissues seldom do more than stimulate or block a preexisting process, much like enzyme induction.

Repression and depression of complex processes offer numerous opportunities for somatic adaptation and for generating nonlethal phenotypic variation of certain perhaps useful forms. The response itself—which is an entire developmental process, whether as massive as the formation of the nervous system or as modest as the formation of a hair follicle—can be extremely complicated and involves multiple cell types and complex cell behaviors. If, however, it is elicited by a single signal, it can be modified in amount and transposed in space and time by merely transposing the signal or by secreting an inhibitor of that signal.

The redeployment of ready-made developmental pathways and cell types under the influence of simple signals obviates the complex task of reestablishing these processes from the component parts. When one thinks of evolutionary change from a regulatory point of view, it may be hard to divide the complexity of a process evenly into two slightly less complex processes, both of which would have to be enabled at the same time and at the same place. It makes more sense to divide the complexity of a process unevenly into a broad and complex response to a simple and local permissive signal. In this way, only one complex process has to be properly regulated, and one simple process properly placed.

How Proteins Do It

At the heart of enzyme induction in bacteria is the repressor, which in response to binding lactose somehow changes its shape, causing it to fall off the DNA. Here we have an example of *signal transduction*, where one kind of signal, the level of a metabolite, is transduced into another type of signal, the binding of a protein to DNA. This form of linkage is widespread in all of biology. In understanding how the repressor responds to lactose, we are solving a much more general problem of how molecules talk to one another and transmit signals.

Jacques Monod's first great insight, derepression, showed how the genome could respond intelligently to simple signals. His second great insight, which he called *allostery*, explained how proteins do the heavy lifting of decision making. Allostery, from the Greek *allo* meaning other, and *stere* meaning solid, referred at first to the fact that proteins can have two kinds of sites of interaction. One is the locus of the protein's function, and the other is the locus of regulation of that function. The protein has a functional part and a regulatory part. Although this seems unexceptional to us now, in 1965 it was a profound insight. It was profound not only because it contradicted the prevailing view of biochemists at the time, that each kind of enzyme had only a single kind of site for carrying out its chemical reaction, but also because it liberated proteins to engage in an unconstrained variety of regulatory interactions. Part of the profundity follows from the fact that at the level of the atomic dimensions of proteins, explanations of their behavior can no longer be vague and ad hoc. They must conform to the laws of chemistry and physics. Allostery was no hand-waving model, but a chemical model that showed how a molecular switch operates. (Monod's conceptual model was published just after the first atomic-level structures of proteins were completed.) In his model permissiveness was explained at the molecular level. These insights prompted Monod to exclaim with characteristic bravado, "I have discovered the second secret of life."[11]

Biochemists at the same time were working on another problem of regulation that involved not control of the synthesis of an enzyme but instead direct control of an enzyme's activity. Enzyme inhibitors were well understood, and several were well-known drugs. For example, sulfa drugs directly inhibit an enzyme used to make a component of DNA, penicillin inhibits an enzyme that makes bacterial cell walls, and the AIDS drug AZT inhibits a viral enzyme used in replicating human immunodeficiency virus.

All of these inhibitors act by impersonating the normal target of the enzyme, known as the substrate. The inhibitor occupies the site on the enzyme where chemical reactions occur and physically blocks the binding of the real substrate. Ever since Emil Fischer drew the analogy in 1894, enzymes and substrates had been compared to locks and keys fitting together. An inhibitor was a false key that fit well enough into the lock to keep other keys from entering, but not well enough to turn the tumblers and open the lock, that is, to undergo a chemical transformation. Thus, inhibitors were expected to bear many likenesses to the substrate.[12]

By 1960 paradoxically several enzymes were already known to be inhibited by molecules that looked nothing like their substrates. Enzymes that stood at the beginning of a biosynthetic pathway were often inhibited by chemical entities produced at the end of the pathway, many steps removed. Hence, the whole process was called feedback inhibition.

Feedback inhibition makes logical sense. If you owned a factory manufacturing automobiles and sales were so sluggish that finished automobiles piled up in the showrooms, you would cut back your purchase of raw materials such as steel, rubber, and glass. It would make no sense merely to slow down the final steps of assembly, such as the paint job; while curtailing the production of finished autos, the process would still consume costly materials, energy, and labor.

Since the end product of a pathway did not generally resemble the substrate used by the enzyme of the first step, it could not impersonate that substrate. Regulatory control had to be exerted at an alternative or "allosteric" site. Commenting on allostery, Francis Crick, codiscoverer of the structure of DNA, said in 1971, "That meant that you could connect any metabolic circuit with any other metabolic circuit, you see, because there was no necessary relation between what was going on at the catalytic site and the control molecule that was coming in."[13] Separate sites were designed independently.

Monod and Jacob were aware that this model went beyond metabolic control and had significance for the evolution of circuits coordinating complex processes and hence for the evolution of complex organisms. Freeing the business side of the enzyme (the catalytic site or primary binding site) from the regulatory side allows their independent evolution, without the constraint imposed on a single site to meet dual functions. The catalytic site is constrained by all of the specialized chemistry of catalysis. The regulatory site can be constructed to interact with almost anything that has regulatory relevance. Since much of evolution involves connecting conserved core processes in new ways, it is a distinct advantage to separate the functional part of a protein from its regulatory part, which can then evolve in an unconstrained manner.

Today it is evident that many proteins are modular, having separate functional and regulatory parts, and that through the regulatory part they communicate signals across very different pathways—from cell proliferation to protein synthesis, from metabolism to heart rate, from inflammation to cell death. The power of proteins to integrate new regulatory connections in a simple way has fueled much of their change during multicellular evolution. Proteins with different domains arise readily in evolution, a fact that creates a major deconstraint on the evolution of regulatory connections.

Although the biological implications of separate domains on the same protein were profound, mechanistic insights awaited an understanding of how the allosteric site could actually control the catalytic site, despite separation from it. Evidence was building for the notion that protein molecules are not rigid and can have more than one folded conformation or shape. Monod's conceptual breakthrough on the mechanism of allostery was to argue that allosteric proteins have not one but two conformations that differ in the degree of activity of the active site. The protein was itself a molecular switch having active and inactive states with regard to enzymatic activity. He postulated that the protein could oscillate freely between the two conformations. It was like a toy that could flip into a new state, where all aspects of the geometry were altered. This was a model that also described the active and inactive forms of hemoglobin, as recounted in Chapter 3. Although not an enzyme, hemoglobin was recognized to be an allosteric protein with a high-affinity oxygen-loading state and a low-affinity oxygen-unloading state.

For allosteric enzymes, the two states differ not only in their catalytic activity but also in their ability to bind the regulator. If the inactive conformation binds the regulator more tightly, the regulator is an allosteric inhibitor; binding it would hold the protein in its inherently inactive state. If the regulator binds more tightly to the active state of the enzyme, it is an allosteric activator; binding the regulator would hold the protein in the active state.

In the case of the repressor, which is an allosteric protein, the form of the protein that binds lactose tightly binds DNA weakly. And the form of the protein that binds DNA tightly binds lactose weakly. Therefore, when lactose is present, the protein is held in the state where it binds DNA poorly; the repressor stays off the DNA, and transcription begins. In the case of hemoglobin, diphosphoglycerate binds to the inactive state and therefore is an allosteric inhibitor, releasing oxygen in the tissues. Allostery was aptly named because it implied a change in the "solid" shape of a protein. Allostery came to mean that the protein had "alternative conformations" or "two states."

Monod was particularly proud of his assertion that in a protein made up of multiple subunits, each with its own enzymatic and regulatory site, all subunits pass concertedly from one conformation to the other. Such organization makes the response behavior of the protein all or none. In this way, the repressor protein's on-off behavior is converted into a transcription on-off behavior. Monod asserted that everything in biology is either all or none, on or off. Intermediate levels of activity reflect a mixed population of molecules, some of which are entirely on and some of which are entirely off—just like the switch on the toy electric train. This simple idea has stood the test of time.

As we move to the molecular level, the allosteric model faithfully maintains the distinction between permissive and instructive signals. In Monod's model, the inhibitor that binds to the regulatory allosteric site does not instruct the enzyme to change from an active to an inactive state; it merely binds preferentially to the preexisting inactive state, encouraging that state to persist and accumulate in the population. The inhibitor is really a selector of a preexisting response, not the creator of a response. Selection is the mechanistic basis of permissiveness. (In Chapter 5, we will consider selectors that choose among an unlimited number of states. Though the choices are more varied, the principle is the same as in allostery.) Much of allosteric enzyme regulation is essentially designed into the protein in advance of the regulator's arrival. It is not that permissive signaling systems are less complicated than instructive ones. It is that permissive *signals* are less complicated, thanks to the complex prepared responses of the receiver. Permissive signaling is weak linkage, because the signal does not alter the actual process; it merely a selects upon it.

We are now better able to understand how conserved and constrained mechanisms facilitate variation around them. An allosteric protein is highly constrained. A typical protein has thousands of weak chemical interactions that collectively hold it in a single stable configuration. It took a great deal of metabolic energy to build the protein, from the synthesis of the amino acids to the synthesis of RNA and its translation. An allosteric protein is poised on a knife-edge, with two stable configurations differing in activity and in how strongly a regulator binds to an allosteric site. The protein shifts continually from one state to the other. The allosteric protein is a design so constrained that it cannot endure mutational change without damage to this allosteric function. Yet this extensive internal constraint enables extensive deconstraint in the evolution of regulatory connections. Allostery makes the protein capable of weak linkage.

A regulatory signal does not have to generate the active or inactive state—those options are already built in. It simply selects one form or the other by binding more tightly to it. The evolution of such regulatory sites has few constraints, for the regulator has little to accomplish. It does not have to interact with the highly constrained and precise catalytic site. The regulatory site can be almost anywhere on the protein surface.

We have delved so deeply into in the workings of this form of physiology because two-state proteins are important well beyond metabolic control. The greatest novelty that has evolved in multicellular organisms is the passage of information, not the chemical rendering of metabolic intermediates. Much of the transfer of information comes from switch-like molecules that can exist in two conformations. For switch-like molecules, the core mechanism is allostery. Such molecules communicate much of the information for control of cell growth and cell differentiation. Shown in Figure 20 is the atomic structure of the switch-like protein Ras, which is defective in its switching in a majority of human cancers, particularly cancer of the pancreas and colon. It exists in two states and communicates information from signals outside the cell to internal pathways leading to cell proliferation.

Allostery promotes weak linkage by separating regulation from function. Continual selection for the retention of weak linkage facilitates the generation of phenotypic variation and deconstrains the selection for new functions and new regulatory connections.

Localization and Recruitment

Signals mediated by allostery can control an enzyme's activity in ways other than by affecting either directly or indirectly the active site of the enzyme. The preoccupation of biochemists with the active site caused them to attribute all specificity of an enzyme to discrimination among substrates at the active site. But there is a simple alternative to

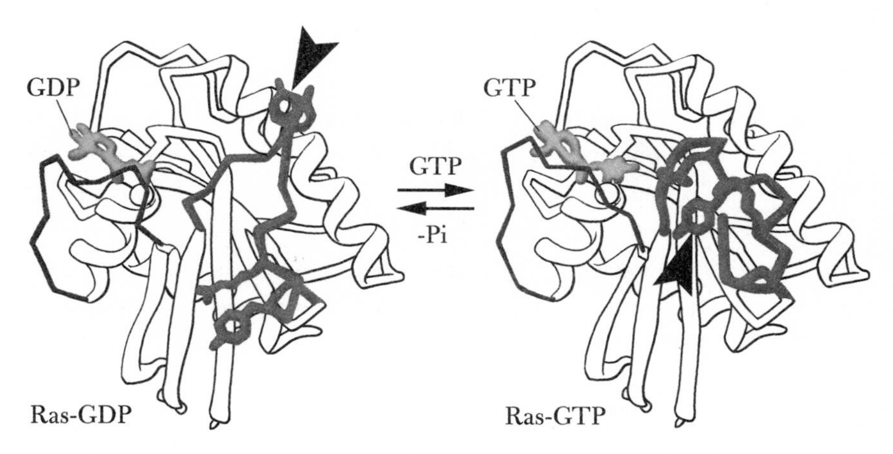

Figure 20 Switch-like proteins. The overall structure of the Ras protein in two conformations: the form that binds GTP and the form that binds GDP. GTP and GDP are energy-rich metabolites of all cells. Note the conformational shift in the folding of the polypeptide chain indicated by the arrowhead. In these two states it interacts differently with signaling proteins.

specificity at the substrate-binding site. Enzymes can be built to be nonspecific, and specificity can be generated by controlling their access to it of potential substrates. An allosteric site can be used to bring a candidate substrate close to the nonspecific active site and exclude irrelevant targets.

Control by localization is widespread in eukaryotic cells, which are highly compartmentalized in comparison to prokaryotic cells. Eukaryotes have many places to put proteins. They have several intracellular membrane-bounded compartments, such as the nucleus and the mitochondrion; they have many regions circumscribed by cytoplasmic filaments, and they have protrusions of the cytoplasm like flagella, cilia, axons, and dendrites.

Frequently, the activity of enzymes is controlled by such localization. They are continually active and indiscriminately modify many proteins at low levels; specificity is achieved by the allosteric sites concentrating the substrate near the enzyme. This activity is deconstraining, because the enzyme during evolution can retain its broad specificity, while the targeting of its effects can be controlled by small binding sequences that determine its localization. Control by proximity can be very readily engineered, for binding sequences may be very simple. By common genetic events such small binding sequences can be introduced into proteins. Highly constrained allosteric conformational transitions are not needed, merely a small leash on the enzyme tying it to another structure. As we have seen, much of the gene regulation in eukaryotic cells is achieved by controlling the proximity of proteins.[14]

Proteins are well suited to bearing multiple sites. They are very large molecules on which the active (enzymatic) site occupies only a fraction of the surface. Most of the surface is a potentially free parking area for binding other proteins and for modifications. There can be numerous weak interactions between two proteins where they bind. This association is relatively easy to engineer by a random change in evolution, as compared, for example, to the human engineering of pharmaceuticals, where specificity and affinity have to be squeezed into a small surface area (proteins are typically a hundred times larger than drugs). Thus, the concept of allostery, or the use of alternate sites for binding and for regulation, has permitted widespread change in evolution and has widened the possibilities for easily engineered activation and inhibition.

Facilitating Evolutionary Change

Transcription and signal propagation are core processes that have been highly conserved during evolution. Yet their regulatory connections are some of the most highly diversified in biology. Every new gene in evolution must somehow be linked to a transcriptional regulatory program, and old genes continue to undergo changes of regulation. Every time an innovation occurs, these processes invariably change. Even seemingly unchanged pathways of development mask continual changes in DNA regulatory sequences. The exact regulatory connectivity is often quite fluid while still leading to the same end, as seen, for example, in genes of different *Drosophila* species. The time and place of expression of particular genes involved in embryonic development are often the same in the different species (which only subtly differ in anatomy), but the sequence of DNA in the regulatory regions of these genes has changed a great deal. Presumably under strong selection, the developmental function can be maintained but the regulatory sequences can change. Of course, if the selective conditions changed, the rapid change in the regulatory sequences could presumably generate a new time and place of expression of the gene, a new phenotype.[15]

Weak linkage explains why complex organisms can function with a relatively small number of highly conserved pathways for transcription and signal transduction, yet maintain an extraordinary capacity for physiological and evolutionary adaptability. In molecular terms, weak linkage has two rather unrelated meanings. In the first, it means that the linkages are easily reconfigured because the physical interactions between components are not unique or highly specified. Proteins interact by stabilizing or enabling an already complex process, rather than by adding key structural elements to change the process fundamentally. The other meaning, also often true, is that the interactions of the proteins with one another or with DNA are energetically weak compared to "strong" structural interactions among proteins like the collagen subunits that make up cartilage. The two meanings of the word "weak" in weak linkage, referring to reconfigurability and to unstable interactions, underlie the permissive and switch-like behavior of many biological processes.

On the negative side, we might expect that where the interactions are not strong they would be extremely error prone, and that errors would be easily propagated in the organism's circuits. Although this is true for individual elements, the circuits themselves are often complex and are built to prevent or correct errors. It is for this reason, we believe, that so many pathways in biology seem to be backed up by other pathways, so-called redundancy. It is a common experience in mouse genetics that deletion of a specific gene, thought to be very important, produces instead a mild phenotype or none at all. Later it is usually found that some other gene's function covers for the deletion.

Why would organisms not build simple pathways of high fidelity rather than complex pathways that are accretions of subpathways of low fidelity? The answer may lie in the retention of weak linkage as a principle on which pathways are constructed. The same pathways are used over and over again within the same organism for different purposes. Thus, they must be modified slightly to interact with a variety of processes and to work in different environments and cell types. Versatility of components and pathways comes at a price. In order for a complex organism to function with low-fidelity circuits, they must be overdetermined and maintain separate circuits that respond to failures.

An example of a signaling pathway is the control of glucose levels. The hormone glucagon is made in the pancreas and released into the blood, where it signals cells of the liver to break down glycogen to glucose, an effect opposite to that of insulin. In this circuit no direct interaction takes place between the glucagon signal and the enzyme that catalyzes breakdown of glycogen. Glucagon does not enter the cell; instead, it binds to a receptor at the cell surface. The receptor, which sticks through the membrane into the cell, activates a switch-like allosteric protein in the cytoplasm, like the Ras protein, which has preexisting active and inactive states. The protein interactions are weak and of low specificity, just enough to stimulate an already poised process. In its active form, the switch-like protein binds another protein, an enzyme, activating it to produce a small molecule inside the cell, called cyclic AMP. This molecule diffuses everywhere and binds to another allosteric enzyme, which in turn modifies a third allosteric enzyme, the one that actually breaks down glycogen. It is a cascade of weak linkages. Such a system is flexible physiologically, because of the many places where other inputs can be received and because the components have their inherent active-inactive states. For example, if increased heart rate requires more glucose, there is another indirect pathway that can independently increase the level of cyclic AMP.

Though it might have been possible to design a pathway for glucagon to travel from the pancreas to the liver, enter the liver cell, and bind to the glycogen-degrading enzyme, stimulating its enzymatic activity directly, such a path actually would be constraining and difficult to engineer, with many precise requirements for location and fit. First, one would need to evolve a special mechanism to transport glucagon across the plasma membrane. Second, it would be hard to design a catalytic site on the enzyme that could accommodate such a dissimilar molecule in addition to glycogen, which is a large substrate. Third, other inputs, such as insulin, affect the activity of this enzyme, and these too would have to be engineered into a very small active site. Finally, and most important, the glucagon/cyclic AMP pathway is used over and over again with minor modifications in different cells and in different organisms. There is an economy in conserving the same versatile circuit: the outputs can be easily varied. That would not be the case if the signal were an intimate part of the response. In a related example, an adrenalin signal uses a common pathway to different ends: in the heart, it increases the force of contraction; in the liver, it increases breakdown of glycogen; in the lungs, it dilates the bronchioles; and in the gut, it decreases motility.

Facile Connections Require Facile Receptivity

Although we have illustrated that weak linkage lowers the barrier for new connections, we have ignored that weak linkage requires a corresponding capacity of systems to be "linkable." In an electrical analogy, it is not sufficient to have compatible plugs and outlets, the appliances must operate on the same voltage and frequency.

Probably the most graphic illustration of "linkability" in biological weak linkage is the adaptability of neurons to different signals. The nerve cell is an ancient cell type dating back, it is thought, to Precambrian jellyfishlike animals. The variety of nerve cells within an organism and between organisms reflects the diverse signals used to transmit signals to other nerve cells or muscle. All nerve cells generate an electrical voltage across their outer membrane. Neurotransmitters secreted by nerve cells bind to receptors on the plasma membrane of other nerve cells (or muscle cells). The nerve cell has many kinds of receptors and ion channels, some admitting positive ions and some admitting negative ions. It acts as a small computer, summing positive and negative ion inputs from each channel. Electrical signals ultimately trigger calcium release, which in turn triggers the secretion of neurotransmitters, allowing the particular nerve to signal to another nerve cell. Channels that raise or lower the membrane voltage contribute in opposite ways. Only when the aggregate voltage gets to a certain level does the neuron fire.

The nerve cell is an exquisite example of weak regulatory linkage functioning in a poised two-state system. Signal and response have no physical linkage. The receptor-ion channels do not touch the secretion apparatus that is ready and waiting at the other end of the cell. It is the electrical impulse traveling the length of the nerve that connects the receptor-ion channels and the secretory apparatus. Given that there is no physical connection and no requirement for exact fit between the receptor ion channel and the distant secretory apparatus, there is little constraint on making new connections.

The nerve cell itself, though, is constrained to change and is conserved in its basic properties and components, namely, all those for generating the membrane potential, impulse propagation, calcium release, and neurotransmitter release. As shown in Figure 21, the entire cell is organized to exist in two states, polarized (secretion off) or depolarized (secretion on). The trade-off for the internal constraint is the regulatory deconstraint of accommodating so many kinds of inputs and outputs. Any of a wide variety of receptors and ion channels can be introduced, and all will work because they all contribute to a common currency, the membrane voltage. In this way, different kinds of nerve cells with different receptors and neurotransmitters can be produced within an organism using the same basic cellular design. The organism varies the signals and responses to generate several types of neurons, and evolution uses the same property to alter these over time.

The genome of multicellular animals is itself set up in a way that facilitates the evolution of new genes, using a kind of weak linkage among all the protein-encoding domains in the genome. During Precambrian metazoan evolution, new genes were created by fusing together various pieces of other genes, especially new genes for components of signaling pathways and of the extracellular matrix. The

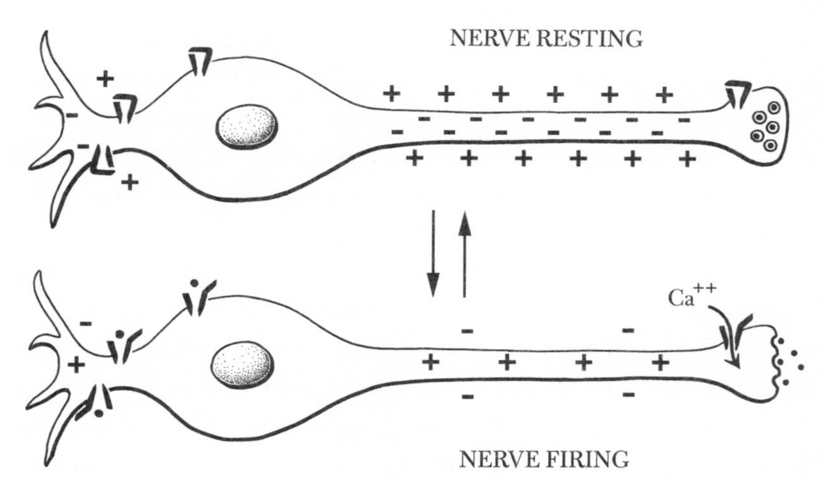

Figure 21 The on-off states of nerve cells. The entire cell is built to exist in a resting state (conveying no nerve impulse) or a firing state. When it receives a threshold level of signal, it fires. Then it reverts to the resting state.

fusion of gene pieces is in principle beset with problems, because fusion must be exact to have all the functional pieces in phase. The problem arises because three bases determine a single amino acid, the triplet code. If the fusion were one base off, all triplets downstream of the junction would be out of phase, giving a completely different and nonfunctional protein.

Given the size of the genome and the large amount of noncoding DNA, one might imagine that the likelihood of making exact couplings of many pieces into one large gene, for one large protein, is small. However, we now know that most genes encode proteins in pieces. Each short coding piece is an island surrounded by long stretches of noncoding DNA, called introns. During transcription, the introns are spliced out precisely, yielding messenger RNAs coding for multidomain proteins.

The machinery for splicing together the RNA is a complex, highly constrained, and extremely conserved aggregate of about two hundred proteins and RNAs. It recognizes general sequence features at the two boundaries of the intron flanked with coding sequences. At these boundaries, it perfectly cuts out the intron and splices together the ends of the remaining RNA. Still, this highly constrained and conserved splicing machine is tolerant of the length of intron sequence between two boundaries. If a new piece of coding sequence with its intron boundaries (and bits of its introns) is placed within an existing intron elsewhere in the genome, it will be spliced properly and incorporated into the final messenger RNA, in the correct frame. Hence its encoded protein domain will be incorporated into final protein.

Since there are several mechanisms for moving blocks of DNA around the genome and for dropping those blocks into existing introns, the structure of the genome with its long intron sequences facilitates the formation of new multidomain protein structures. Thanks to the RNA splicing machinery, the sequences can be dropped in without precision and still incorporated precisely into new structures. The large expansion of multidomain proteins that coincided with and presumably supported the elaboration of multicellular organisms 600 million years ago, made excellent use of the organization of the genome into coding and intron sequences and of the effectiveness of the RNA splicing machinery.

Weak Linkage and Evolution

The selection for a small number of conserved core processes versatile enough to be used in many different contexts to support the complexity of large multicellular organisms is a product of selection for physiological adaptability. As a side effect, core processes with high adaptability have a high capacity for weak linkage. Such processes are responsive to genetic changes of regulation. They have been used in many different combinations at many different times and places in the organism's development and physiology, so it is likely that processes capable of weak linkage pose little barrier to future use in different combinations, times, places, and amounts.

The capacity for weak linkage, which is built into the processes, is reselected with physiological adaptability, and thus is a conserved property. Ancient processes such as signal transduction and transcription are constrained to change, but they readily allow regulatory changes that alter the interaction of the conserved components with inputs and outputs. This freedom to link inputs and outputs is a significant form of deconstraint.

Much of the skepticism over the years about the capacity of random mutation or genetic reassortment to generate phenotypic change has arisen from the assumption that genetic changes must create very specific, multiple, complex phenotypic changes. Our view is that specificity and complexity are already built into the conserved processes, as is the propensity for regulatory coupling. It is not necessary for genetic change to create those characteristics, though they are still needed for heritable change.

While the targets of genetic change are not fully known, they certainly include not only the regulatory DNA regions of genes, but also the small inhibitory RNAs, the regulatory parts of proteins, including protein phosphorylation and protein degradation signals, translational control and splicing sequences, and perhaps a host of protein modulators of signaling pathways. Regulation is diverse and pervasive; there are a multitude of targets. These targets are easily connected to one another by weak linkage, requiring only small mutational change. Both the capacity for weak linkage and the receptivity to weak linkage are highly selected traits in organisms and are deeply conserved.

FIVE

Exploratory Behavior

We have seen that existing somatic adaptations can be a ready and available source for new variation when genetic change stabilizes adaptive processes at different points along their ranges. This application of the Baldwin effect has not been widely endorsed as a panacea for explaining novelty in evolution, because the kind of variation that seems most interesting in evolution is not that which causes small quantitative perturbations of existing systems. Certainly, in anatomical novelty it is hard to imagine how an organism could store within itself the capacity for forming novel structures in the future, such as the first wing or the first eye.

What kind of novelty *might* be stored in an organism, to be stabilized by mutation, for generating new anatomical structures, or for that matter new physiologies or new behaviors? If we shift our attention away from existing highly integrated physiological processes, like adaptation to heat or adaptation to changes in the food supply, and turn to the conserved core processes that underlie physiological processes, we find that their adaptive ranges are very large. The overall phenotypic variation that could be produced if each were allowed to vary over its entire range is much greater than the normal adaptive physiological range of an animal.

A subset of these core processes is what we call *exploratory processes*, or core processes that display exploratory behavior. Their adaptability is central to their function. Examples are found at many levels within the organism, from the subcellular to the behavioral.[1]

Exploratory behaviors are a special and powerful form of somatic adaptation. They generate many, if not an unlimited number of, specific states in the course of their function, and provide a mechanism for selecting among these states those that best meet the particular physiological need. Because they produce so many states, the cell will not use most of them; but under new selective conditions these can generate novel structures. To revert to the toolbox analogy, if we had to tighten a nut of unusual size, we might reach for an adjustable wrench, analogous to a continuously varying somatic adaptation. But if our toolbox were based on exploratory principles, it would magically generate a nearly infinite set of fixed-caliper wrenches of all conceivable sizes, from which we could choose the appropriate one. Although exploratory processes make use of weak linkage, they use it to choose a small number of alternatives from a large number of possibilities. Cellular processes making use of both weak linkage and exploratory mechanisms play a major role in facilitating evolutionary change.

We give several examples here of exploratory mechanisms that demonstrate not only the broad range of circumstances in which they are used in biology, but also the mechanistic diversity of exploratory processes. All mechanisms share the property of generating variation that is completely random and constrained very little, followed by functional selection among the diverse states. This is variation and selection in the physiological domain.

When the cytoskeleton is visualized in most cells (except in unusual cases where it is highly organized, as in striated muscle), it seems to be a chaotic collection of filaments, which nevertheless display an overall organizational bias related to cell shape. We know from experiment that these filaments are not merely adapting to the shape of the cell but are actually producing it. The cytoskeleton is highly plastic and is responsive to external stimuli and signals internal to the cell. Biochemical studies have shown that one of the major components of the cytoskeleton is organized by trial and error. Such a process involves the continuous generation of randomly oriented filaments, followed by selective stabilization of those filaments that reinforce the shape of the cell compatible with its function. This kind of process gives cells an almost unlimited capacity for variation in their anatomical organization and a wide responsiveness of that organization to external signals.

On a different level of function, trial and error plays a major role in the behavior of an entire organism. For example, when we look at how ants forage for food and learn to exploit their discoveries, we find a trial-and-error strategy that bears an eerie resemblance to the means by which the cytoskeleton self-organizes.

Evolutionary biologists were not persuaded that novelty can arise merely by stabilizing existing somatic variation, particularly in the realm of anatomical change. What in nature could foreshadow the human brain with its high cognitive capacity? The development of the nervous system raises this question, as well as a related question of how human beings with only 22,500 genes can specify trillions of cells and synaptic connections. One might ask where in the genome is the complexity of the contemporary organism encoded, even before one asks where in the genome cryptic future adaptations are hidden.

Some of the answers to both the present and future complexity of multicellular organization are found in exploratory processes based on randomness and functional selection. The nervous system thereby can construct itself with a relatively small number of rules. The plasticity afforded by physiological variation and selection not only accounts for much of how the organism generates the complexity of the nervous system or other forms of anatomy, but also for how these systems repair damage (as in recovery from injury and stroke). It also helps explain how new anatomies can evolve from existing forms. Exploratory behavior is especially evident in conserved processes operating in the spatial dimension.

The processes for generating physiological variation and selection are themselves complicated. In some cases, the organism goes to great lengths to generate variation at each phase of a physiological adaptation. This capability is seen most clearly in the vertebrate adaptive immune system, which is based on exploratory principles. (We do not discuss it here, since we wish to follow the path toward spatial and anatomical organization.) In other cases, a certain diversity of outcome is the result of a broadly receptive process that generates limited variability and randomness.

If we were to ask how the organism could build a vascular system that delivers oxygen and nutrients to every cell in the body, controlling the delivery so that tissues with strong demand get more oxygen and nutrients than those with small demands, we would be describing a system of considerably greater complexity than the interstate highway system in the United States, and a system that does not suffer rush-hour congestion when demand increases. The vascular system uses exploratory mechanisms to respond to local needs. It generates limited variation and achieves its final structure by selective stabilization. Such a system can grow with the individual, can vary to meet demand, and can easily change during evolution.

It is the evolutionary role of exploratory processes that causes us to give them such prominence in this book. They seem to be able to overcome barriers to novelty, since they generate novel structures in the course of their normal physiological function. This competency addresses the problem of evolutionary adaptations that require simultaneous events. The eye was the classic problem that defied explanation, burdened as it is with the requirement that so much must go right simultaneously to produce even the most minimally functional organ on which selection might act. The problem would be solved, Darwin thought, if there were "numerous gradations from a simple and imperfect eye to one complex and perfect . . . , each grade being useful to its possessor . . . Then the difficulty of believing that a perfect and complex eye could be formed by natural selection, though insuperable by our imagination, should not be considered as subversive of the theory." Yet might it stretch credulity to have so many independent events, each with *no selective value*, to form the first simple eye, the first wing, the first lung, or the first placenta? Might processes that generate significant variation in their routine function also reduce drastically the number of steps to achieve novelty?[2]

At the end of this chapter we specifically discuss the role of exploratory processes in overcoming the requirement for simultaneous change. We have chosen to examine the vertebrate limb. The limb, like the eye, has a complex anatomy and many cell types. The plausibility of the rapid evolution of the bat wing, for example, hinges on whether at each stage of modification a substantial enough improvement took place for selection to be effective. In the case of a wing, the initial improvements from the mammalian forelimb to establish gliding or flight might be substantial. (To appreciate the problem, bats did not evolve from flying squirrels or, as once thought, from flying lemurs; it is now believed that bats are most closely related to whales, dogs, and deer.) But how many features would have had to change at once?

Limb evolution entails simultaneous change in many tissues: innovations in bone or cartilage anatomy, positioning of the muscles relative to the bone, innervation of the new muscles with nerve cells originating at great distances, and provision of a new balance of nutrients and oxygen through the vascular system. We shall see that the inherent somatic adaptability of these systems through exploratory mechanisms drastically reduces the barrier to novelty. At each stage an exploratory process could, even without genetic modification, adapt to changes in anatomy. Such highly adaptive processes facilitate the production of significant viable and novel variation on which selection can act.

Variation and selection in another context of course underlie the Darwinian model of evolution, whereby the organism as a whole generates extensive heritable phenotypic variation and the fittest of the variations is stabilized by selection. The parallel between evolutionary and somatic adaptation is not superficial, but goes to the heart of why exploratory systems make such an important contribution to evolution. In addition to a role in decreasing lethality (by reducing collateral damage from other changes in an organism), somatic adaptations involving exploratory processes offer many targets where genetic change can substitute for somatic change. This is another instance of the interchangeability of genetic change and environmental change, as we discussed for sex determination and hemoglobin in Chapter 3.

Exploratory systems, which are broadly responsive, generate many states, any of which can be stabilized by peripheral signals. Typically, the exploratory processes are highly conserved; it is the stabilizing signals that change. When generated by the organism, these signals can be altered by genetic reassortment or mutation, which changes the selection on the exploratory system but leaves unchanged its capacity to generate multiple states. Thus by a change in the selective signal, a cell can be selected to crawl along any path, an axon can be selected to sprout from any position, and the vasculature can be selected to ramify in any direction. The large number of possible somatic states is easily transformed into a large number of possible phenotypes on which selection can act. This reduces the requirement that selection stabilize only very minor modifications. Substantial phenotypic variation comes out of the many somatic variations generated in the course of the normal function of the organism.

How Cells Get Their Shape

The architecture of cells is achieved without an architect. No central regulation is discernible. Cells are in fact capable of many structures; many are chameleons that change their structure in response to circumstance. The free-living *Amoeba proteus* was aptly named for the Greek sea god Proteus, who could transform himself into any shape. This capability creates a huge reservoir of somatic adaptation for cells, which becomes a substrate for evolutionary change, much of it based on exploratory principles.

The proteins used to generate cell shape are like all other proteins encoded in the cell's DNA. Although DNA sequences control the time and circumstances of expression, DNA provides no instruction on where to place the proteins in the cell. There is no genetic information for large-scale cellular organization. Furthermore, cells having the same DNA and inhabiting the same environment can have very different shapes. Cell shape responds to developmental and environmental cues independently of genetic control. The capacity to change shape underlies significant processes, such as the directed migration of cells into the margin of a wound for repair, the extension of nerve axons to different targets in the development of the nervous system, the contortions of a white blood cell when it engulfs a large particle or when it infiltrates the lining of a blood vessel to hunt down an infection, and the extensive remodeling that occurs in each cell during cell division. Part of the process of achieving cell organization relies on trial and error, a form of physiological variation and selection at the level of protein assembly.

To understand how adaptability of cell shape provides opportunities for evolutionary change, we must descend once again to the molecular level. It was not until the 1970s that the skeletal elements of the cell were revealed, and it took another decade before their mechanistic properties were understood. Cells have an internal skeleton, called the cytoskeleton, made up of three different families of long, thin filaments. These filaments criss-cross the cell interior in arrays reflecting the different cell shapes. Each type of filament is composed of its own kind of globular protein unit; in each filament a hundred thousand or more identical globular protein units may be linearly assembled. The mitotic spindle, whose fibers were perceived by early microscopists to connect the chromosomes to the poles at cell division, is made up of microtubules, which are also widely used in the cytoskeleton of nondividing cells, such as nerve cells. The other two major filament types, actin and intermediate filaments, along with microtubules, play structural roles in the cell—a different role for each. The cytoskeleton is both rigid, giving any specific cell stability against mechanical deformation, and versatile, capable of being reassembled and used over and over again to support different shapes.

The key to the adaptability of microtubules is their dynamics. In a typical nondividing cell, hundreds of microtubules radiate out toward the cell membrane from a central nucleating structure. In this configuration the microtubules, like spokes of a wheel, seem to give rigidity to the roughly polygonal cell.

Yet this initial characterization of microtubules, as rigid rods giving a cell its shape, was misleading. It was an impression gleaned from the early fixed histological preparations of cells. (In similar vein, a single snapshot of a football game, as opposed to a movie, would also give a misleading impression of the event.) When specialized methods allowed movies of microtubules to be made within living cells, the microtubules proved to be dynamic. They continually grow, disintegrate, and regrow, each individual microtubule persisting for only five minutes. When a single microtubule grows out for a period, then spontaneously shrinks back toward its point of origin, it is replaced by a new one—which grows in a different direction. Over time, the number and nearly random distribution of the population remains about the same, although individual microtubules change.

The entire process of microtubule growth and shrinkage requires energy. This requirement was initially puzzling because much more complicated structures such as viruses assemble spontaneously. In actuality, energy is not involved in assembling the microtubules but instead causes them to disassemble and keeps the turnover dynamic.

The purpose of the turnover was initially unclear. This process, now called dynamic instability, seemed to amount to nothing more than a futile cycle of growth and shrinkage of individual filaments, without changing their overall distribution. We had glimpsed physiological variation without selection; but when selection was included, it revealed a new and powerful mechanism of somatic adaptation.[3]

The function of dynamic instability lies not in the assembly of the individual microtubules, but in the capacity to organize them in arrays. Microtubules extend randomly from their tips and depolymerize back to the nucleation center by loss from their tips, as depicted in Figure 22. They continue doing so until they encounter a stabilizing activity in the cell periphery, which blocks depolymerization at the tip, far from the site of nucleation. Microtubules that randomly enter the region of stabilization persist, while rapid turnover eventually eliminates the others. A particular polarized or asymmetric array is achieved stepwise by local stabilization. In the end, most microtubules extend from the center to the stabilizing region at the periphery. When the cell structure is finally achieved, the dynamics of the microtubules may be reduced, and the arrangement will be more permanently stabilized.

Thus rapid turnover (variation) and local stabilization (selection) can transform an irregular dynamic array into a polarized stable one. The adaptability of this process is such that stabilizing signals can

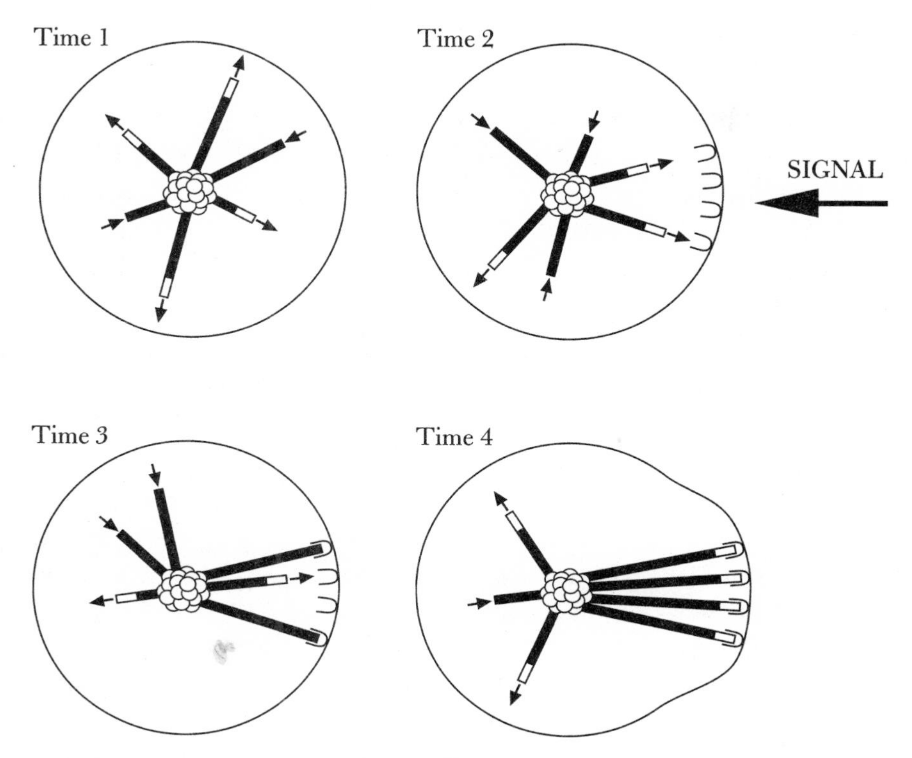

Figure 22 Microtubule exploration. All eukaryotic cells contain microtubules that vary in arrangement under different conditions. Each microtubule grows (arrow out) and shrinks (arrow in) from one end (time 1). When a signal arrives (time 2), stabilizing agents are activated on one side of the cell. Microtubules reaching those agents by chance are stabilized (times 3 and 4) and do not shrink. The final arrangement of persisting microtubules depends on the distribution of those stabilizing agents.

come from any direction, and the microtubule array and the cell will respond appropriately. This mechanism is avowedly selective, rather than instructive. There is no evidence of instruction from external signals directly causing the microtubules to polymerize in specific directions.

The adaptive nature of arrays of microtubules allows them to function, using the same proteins and the same rules, in circumstances as different as mitosis (where a bipolar spindle is formed) and nerve axon formation (where the array is very long and monopolar). In mitosis, chromosomes condense and are left scattered around the cell. In every cell the chromosome arrangement is different, and in different organisms the number of chromosomes is different. When a microtubule's growing tip fortuitously hits a specialized region of the chromosome (the centromere), it is captured and stabilized. Captured microtubules then serve as a scaffold to drag the chromosome to the center of the cell.

In mitosis forces other than microtubule dynamics and stabilization contribute to the formation of the mitotic spindle. Nevertheless, a large component of mitotic spindle assembly and function is the capture and stabilization by the chromosomes of those rare microtubules that interact fortuitously with them. This strategy makes mitosis very robust. The number and initial location of chromosomes can vary, and still there will be a functional outcome. This is critical for evolution, because the adaptability of the cytoskeleton in general accommodates both environmental change and internal genetic change and works in many circumstances.

Cytoskeleton formation is an exploratory process; many potential cell shapes are generated from a single genotype, even under stable environmental conditions. The cell can adapt to any signal that stabilizes any of its numerous potential phenotypes. The mechanism of microtubule turnover does not determine the resulting arrangement of microtubules. Instead, cell organization is driven by stabilizing agents acting peripherally in the broadly responsive and unbiased process of microtubule assembly.

The dynamic cytoskeleton, seen as a conserved core process, facilitates evolutionary change by repeatedly generating new morphologies in two ways: it can be stabilized by extrinsic factors placed by evolutionary change in virtually any location, and it can reduce the lethality of random variation caused by environmental perturbation or by adaptive change in other processes. The capacity for somatic variation and the robust buffering of stress go hand in hand. Since novel phenotypes are included in the broad adaptability of the cytoskeleton, the cytoskeleton supports novelty in evolution. The principles of variation and selection, so powerful a metaphor for evolution itself, are widely employed in many conserved core processes including behavior.

Variation and Selection in Behavior

Of the more stereotyped behaviors, ant foraging is particularly amenable to quantitative study and can be analyzed in simple terms. Ants explore unfamiliar terrain or familiar terrain in which the distribution of food is constantly changing. Thus, experience with their local environment may not help them very much. Ants emerging from their nest cannot see food or smell it. There are no clues to where the food may be; it could be a seed that just blew into the territory. As in microtubules, instructive processes, based on perceiving the food and deriving a strategy of retrieving it, may be difficult to construct. To exploit food, ants need not only to find it but also to communicate their findings to the entire colony. Optimization of foraging could be the principal determinant of an ant colony's success.

The simplest foraging strategy is diffuse foraging. Individual ants go out and return to the nest with or without food, without communicating. For the colony as a whole, advantage would accrue if individuals varied their paths taken, so as to increase the chance of finding food. Variation is important in this sense. However, in this strategy, which several ant species use (so presumably it works for them), there is no selection step and therefore no improvement in exploiting the food once it is found.[4]

A more powerful foraging strategy, based on minimal communication and highly adaptive individual behavior, couples behavioral variation and selection. In this strategy, random exploration by individual ants leads to chance discovery of a food item, which leads to recruitment of other ants, which leads to a more economical and efficient exploitation of the discovered food, as shown in Figure 23.

Ants emerge from the nest and explore at random. As they go, they secrete a highly volatile odorant or pheromone, leaving a temporary trail whose scent they follow to return to the nest. After a

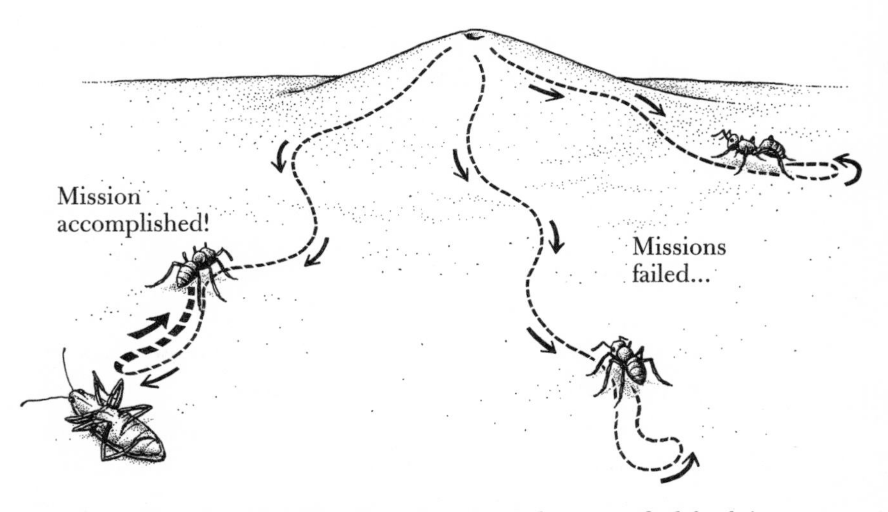

Figure 23 Ant foraging. After an exploring ant chances to find food, it returns to the colony by the same trail, depositing more pheromone signal as it returns. Other ants leaving the colony follow the reinforced trail to the food rather than starting new trails.

period of unsuccessful searching, each ant traces its odorant trail back to the nest. If a foraging ant finds food, it is programmed to secrete more pheromone, reinforcing the trail, as it returns to the nest.

Ants emerging from the nest tend to follow existing pheromone trails, but trails that are not reinforced soon lose their scent. Successful trails become more and more reinforced, and ants emerging from the nest tend to follow those trails leading to food. One problem with this strategy is that it focuses the colony on the first food that is found rather than on the best food. For that reason, the response of ants is not completely deterministic and some level of variation always remains. Even with strong pheromone trails, a few ants will leave the trail to explore anew—a useful lesson for us all.

Ant trail patterns reflect the environment and distribution of food. Different ant species may appear to have different foraging behaviors, because their trail patterns are different. On closer inspection, however, these "species" differences reflect nothing intrinsic about the ant, but instead suggest the different environment or evolved prey preferences of the ants, which in turn reflect differences in food distribution or preferences; the rules of individual behavior are the same: the processes of pheromone secretion, tracking pheromone trails, reversal of path, reinforcement, and random deviation are built into the genetic makeup of the ant. From these rigid rules emerges a highly adaptive strategy applicable to changing environments.

Microtubule assembly and ant foraging are conceptually analogous. Both are exploratory processes involving variation and somatic selection. Ants and microtubules move out in random directions and return if they do not encounter a "target." If they do encounter a target (a stabilizing agent for microtubules or food for ants), the array of microtubules or distribution of individual ants will be modified by a selective process alone.

In the case of microtubules, the ends of certain microtubules are prevented from shrinking back. Other microtubules turn over, while a few each time are recruited and stabilized; as a result, the entire array becomes redirected to a new configuration. The individual ants by their secretion of pheromone rally new ants to follow a specific route. Gradual recruitment occurs without individual ants having to change the means for making their own choices. In both cases change is seen at the level of the population and not the individual unit. This somatic selection is very similar to Darwinian selection, where the environment does not change the individual directly but merely biases the population of individuals.

Microtubule arrays, like populations of ants, are adaptive. The stabilizing signal (or food) does not have to be reliably in the same place. The process tolerates errors. Microtubules can work in many cell types and ants in many environments. The final distribution of stabilized microtubules reflects the distribution of peripheral signals, not changes in the core process of polymerization, even as the distribution of ant trails reflects the distribution of food, not different search strategies. Exploratory variation and selection together are powerful tools for generating physiologies and behaviors that are not merely extensions of existing behaviors. In both processes what the genome encodes is the means to explore, not the outcome of the exploration.

Too Few Genes

The small number of genes in multicellular organisms—14,000 in *Drosophila* and 22,500 in humans—and their high degree of conservation raise two concerns for understanding biology. The first is how the staggering complexity of animals, reaching a kind of apotheosis in the human central nervous system, can be generated from such a small number of gene products. To put things into perspective, the number of neurons in the human brain is estimated to be a hundred billion, and the total number of synapses to be a million billion. They are arranged and function in complex spatial networks. A second concern is that many of the small number of genes are highly conserved; how can the relatively few differences support the extraordinary diversity of anatomy and physiology of organisms on this planet?

The answer to both concerns must come from the use of these genes in combinations. Combinations add up quickly; even 20 different factors deployed in all possible combinations add up to far more than a million billion.

To say that combinations of genes answer the problem of complexity is also to avoid the problem. Even if the cell had a million billion responses, how could there be a million billion signals for these complex responses? Exploratory systems, based on simple rules of interaction, might provide a way to generate complex signals and respond to them with simple functional responses.

Although it is premature for us to claim to understand the development of the human brain, progress in this area of neurobiology and developmental biology has been remarkable. We know that patterning the brain is a mixture of instructive and selective interactions. The number of signals is relatively small, but they are used at different times and different places in generating the gross morphology of the central nervous system. Beyond instruction, there is a significant role for exploratory processes in wiring up the nervous system.[5]

Evolution is about life and death, Malthusian growth and survival. The same kind of life and death selection plays out on the cellular level as part of the process of embryonic development. It is now evident that cell death is a standard part of the program of wiring up the nervous system and that the process of cell death is a stereotyped and highly regulated form of cell suicide. A capacity for suicide, inherent in all cells of multicellular animals, normally is suppressed. In embryonic systems outside of the nervous system, cell suicide is used regionally to generate new anatomies; for example, to sculpt the digits in the hands and feet by causing death of the cells between the digits. The webbing in the duck's foot is a selected default state where the cells are not removed by cell suicide. Cell death also removes transitory structures such as certain early kidney ducts and tubules or the tadpole tail during metamorphosis.[6]

In the nervous system, cell death prunes away superfluous nerve cells after an exploratory process has made tentative connection with potential targets. Cell suicide is the default state, occurring if the neurons have not entered the right locations and made specific contacts.

How is suicide prevented when cells make the right connections? In the 1950s Victor Hamburger and Rita Levi-Montalcini found that neurons needed a *survival factor*, produced by the target tissues to prevent cell death. Neurons that grew into areas that did not provide this factor died. We have learned that the central nervous system, both the brain and spinal cord, produces far more neurons than are ultimately needed for the nerve connections to the peripheral targets. These cells extend their long, thin axons somewhat randomly into the periphery of the body, like foraging ants. If an axon tip by chance enters the anatomically appropriate region, it receives survival factor produced there by target tissues, and it persists. If it enters the wrong region, it receives no survival factor and commits suicide. Since there is a limited amount of survival factor even in the appropriate region, competition and selection occur among neurons.[7]

Once again somatic adaptability facilitates evolutionary change and recapitulates the basic process of variation and selection in evolution. Its potential relevance to evolution has been tested experimentally. If an additional limb, still in its early stages of development, is grafted onto the flank of a chick embryo (midway between the forelimb and the hindlimb), neurons send out axons from the nearby spinal cord region of the host and innervate both the normal limbs and the fifth limb. Since the extra limb enlarges the field of cells producing the survival factor, more neurons survive at the midflank level, whereas they would normally die for lack of a target.

Though new limbs may be uncommon in evolution, the relative size and placement of limbs varies considerably; for example, the very different size of the kangaroo forelimbs and hindlimbs. Experiments show that the process of innervating the limbs can occur without reengineering the process of neuronal growth and survival. At least in the initial stages of evolution, limb evolution could be largely unlinked from the evolution of the nervous system that controls them. The superfluous production of neurons and their peripheral stabilization or elimination can therefore accommodate differences in neuronal targets, adding robustness to the normal development and opening new opportunities in evolution.

Proper Liaisons

The fine patterning of neuronal connections depends on exploratory processes. It is the liaisons between neurons and between neurons and muscle that are likely to be a major target for respecification in evolution. In normal development the cell bodies of neurons involved in motor control reside in the spinal cord and send long axons to the muscle in the periphery. The axon of the sciatic nerve (whose cell body is in the lower back and whose axons reach the big toe) is the longest axon, over 3 feet (1 m) in length. In the periphery, axons often ramify into a number of fine branches that initially contact several different target muscle cells. Eventually, a single skeletal muscle cell is contacted by just one axon.

The initial promiscuous liaisons between nerve and muscle cells must be pruned down to a simple monogamous connection. The choice is not predetermined. When nerve axons grow into the periphery of the body, they are not aware of whether or not a muscle cell is already connected to another axon branch. They are not given a

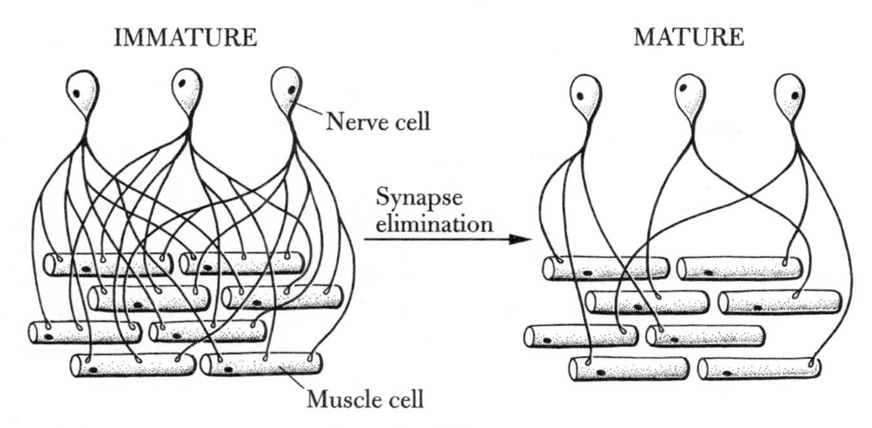

Figure 24 Pruning the nerve terminal. As the vertebrate embryo develops, each nerve cell of the nerve cord extends several axons to the muscle cells. At first, several axon tips contact every muscle cell, shown on the left. As time goes on, the excess axons are pruned, until each muscle cell is contacted by only one tip (shown on the right). Pruning decisions are based on the functional efficacy of the contacts.

specific address, only a neighborhood to enter. The final state is worked out by competition among the various neurons that crowd the synaptic region of a given muscle cell. It is a functional competition, where each neuron fires an electrical signal and the muscle responds, testing the relative strength of synaptic communication. The neuron branch tip with the strongest signal wins out, but as other axons withdraw their branch tips, they redeploy their resources to other targets, as shown in Figure 24.[8]

There are many examples of plasticity of the nervous system based on variation and selection, and increasing evidence exists that refinements in the connections are dependent on functional feedback between the nerve cell and the target cell. The functional interactions can be conditioned by experience; the nervous system responds in an adaptive way, making changes that will affect the individual and not the offspring. Alternatively, the nervous system may respond to hereditary changes such as the growth or disappearance of muscles, thus facilitating evolutionary innovations or modifications.

The plasticity of the nervous system suggests that the initial adaptations to genetic change would not have required modification of the rules for wiring it. The nervous system adapts to changes in the periphery, whether they arise by the normal variability of developmental processes, by damage, by experimental manipulation, by other environmental influences, or by genetic changes that result in evolutionary modification. Evolutionary changes are different in that, after the initial adaptation, time is available to accumulate further genetic changes that can stabilize the modifications—yet another example of the interchangeability of genetic and environmental factors. The basic processes of wiring the nervous system can be conserved, for they are built in such a way that the connectivity of the nervous system can change as the anatomy of the organism changes, without the need to alter the processes that generate that connectivity.

Remodeling Inputs and Outputs

The more we know about biology, the more serious seems Darwin's concern about the requirement for simultaneous changes to establish enough function to allow selection to act on novel structures. Related to this problem is the concern that any change in a system, unless accompanied by balanced changes in other processes in the organism, will most likely create a system less fit than the original system. At least, that is our experience with mechanical evolution. Increase the number of cylinders on the car from four to six and numerous modifications are needed to make those additional cylinders functional. Most important, without simultaneous modifications in the crankshaft, the oiling and cooling systems, and the ignition system, the car would run poorly, if at all. Darwin worried about the lens and the retina of the eye. He was unaware of the complex circuitry for visual processing in the brain, which would seem to render evolution of the eye much more difficult.

The brain and the neurons in the periphery must be coordinated. Otherwise, changes of anatomy in the periphery, which in turn change the pattern of sensory receptors, will not be properly represented in the brain, which has to interpret those sensations. The brain and the body surface do not develop together in the embryo. One would not expect that the early development of the two tissues would be coordinated so that tactile sensation would be accurately represented in the brain. How the brain represents the spatial organization of the body is emblematic of the problem of how to evolve phenotypes in which the various cellular processes, cells, and tissues are coordinated. We find this problem vividly represented in the development of the vibrissae (whiskers) of the mouse.

Whiskers are a vital sense organ that enables the mouse to navigate close spaces in the dark. In the higher centers of the brain are anatomically discrete structures called barrels, which contain masses of neurons. Each barrel corresponds to one whisker on the face, a topographic anatomical representation in the brain of the face of the mouse. Each barrel is a condensation or grouping of about twenty-five hundred neurons, organized roughly as a column going through five layers in the cerebral cortex. The barrel is approximately 0.01 to 0.02 inch (0.3 to 0.5 mm) in diameter.

The whisker neurons do not connect directly to the cortex; there are, in fact, a series of neuronal structures in the relay path from the whiskers to the cerebral cortex. Each whisker is innervated by many nerve endings that connect first to parts of the brain stem called the trigeminal nuclei. Neurons of the trigeminal nuclei send axons into the thalamus, a relay center in the brain for information entering the cerebral cortex. In the cortex, the nerves coming from the thalamus are organized into the barrels. In each of these intermediary structures is a coherent topographic representation of the whiskers: in the trigeminal nuclei of the brain stem where they form condensations called barrelettes, and in the thalamus where the corresponding structures are called barreloids (see Figure 25). Detailed electrical mapping studies have shown that there is a one-to-one correspondence between the nerves surrounding a single whisker and the nerves of a single barrel in the cortex; if you tickle one whisker, a single barrel responds electrically, as does the corresponding barrelette and barreloid in its path.[9]

This accurate topographic map raises interesting questions in de-

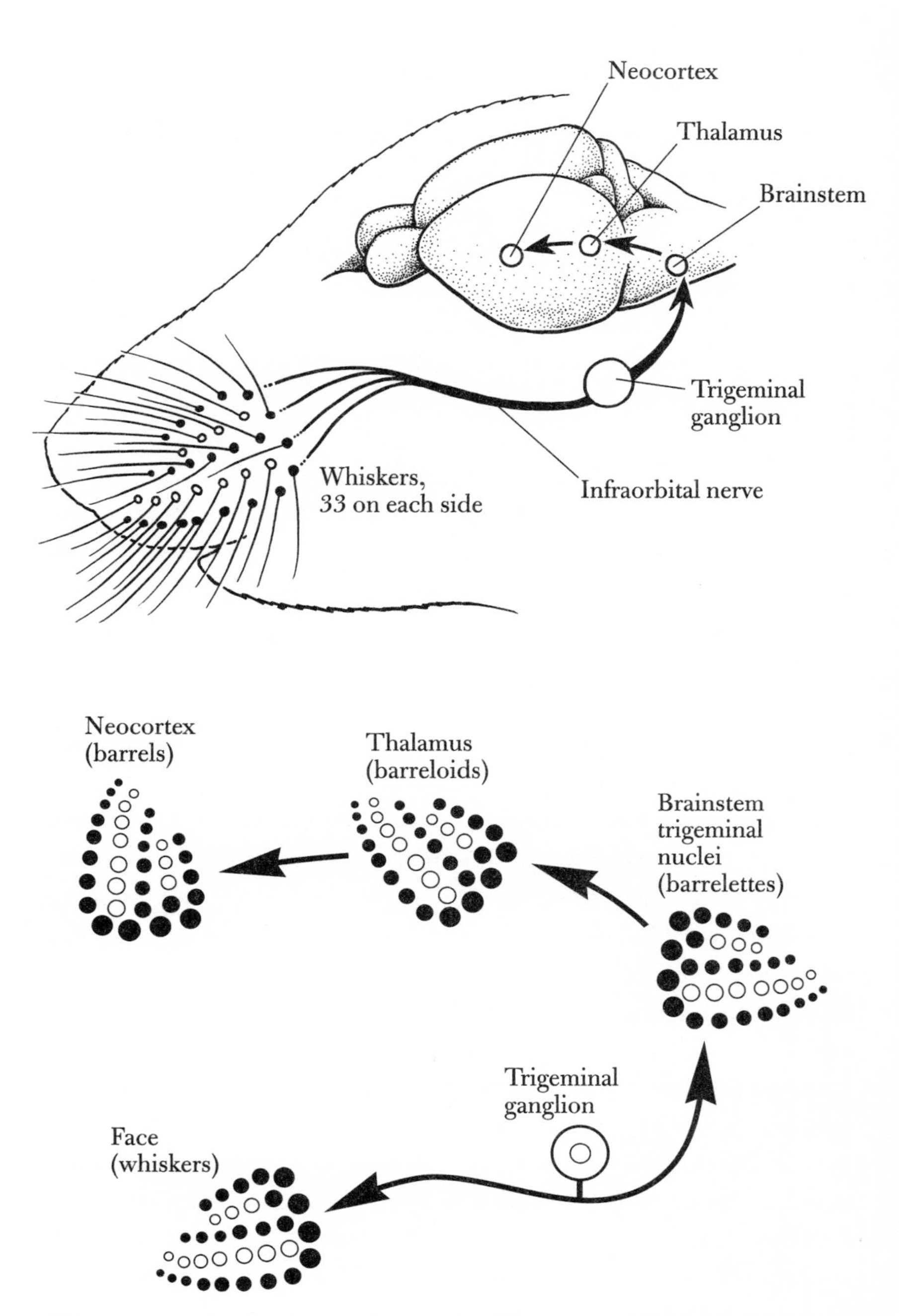

Figure 25 Mouse whiskers and brain development. A normal mouse develops 33 whiskers on each side of its face. Nerves from each whisker enter the developing brain, reaching the neocortex via two relay stations. The neocortex responds to the nerves by differentiating 33 nerve clusters, the barrels, with the same spatial arrangement as the whiskers, rotated by a quarter turn. The relay stations at the brain stem and thalamus form nerve clusters called barrelettes and barreloids.

velopmental anatomy, especially when we realize that strains of mice with different numbers of whiskers have a correspondingly different number of barreloids, barrelettes, and barrels. How are these anatomical structures so closely correlated? Is the developmental process so accurate that each structure arises independently and then they all connect precisely, in the way a engineers build a tunnel through a mountain, starting from both sides and expecting to meet precisely in the middle? Or does one structure form first and instruct the next to form in relation to it? If so, what is the form of the instruction?

Barrels form in the mouse between birth and day five after birth and they remain stable after that time. Some indication of how they form comes from the observation that damage to the whisker region of the mouse pup causes disruption of the barrels in the cortex. After this critical phase, damage in adult animals is less severe but still observable, suggesting that continued innervation from the whiskers is required to maintain the barrel structure.

Trimming some whiskers causes elimination of their barrel domains in the cortex and enlargement of other barrels to take up the space. Removing all whiskers but one leads to reduction in the number of barrels and expansion of the remaining functional whisker barrel to accommodate almost all the original space in the cortex. Trimming one whisker does not destroy the nerves; it just reduces their activity, since the shaved whisker does not get tweaked as much when the mouse navigates in the dark.

These simple experiments demonstrate the key role in controlling brain organization played by exciting nerves in the periphery by whisker movement, and the adaptability of brain development to the inputs. They give us a striking example of behavior (tickling the whisker) altering anatomy (the microscopic neuroanatomy).

Other mammalian examples can be found of barrel-like structures in the cortex that map one to one to a respective sensory field. The star-nosed mole is a grotesque animal whose nostrils each carry a large appendage having 11 fleshy finger-like structures, depicted in Figure 26. Each finger has about a hundred thousand nerve fibers and, like the mouse, the mole uses its tactile appendage to find its way in the

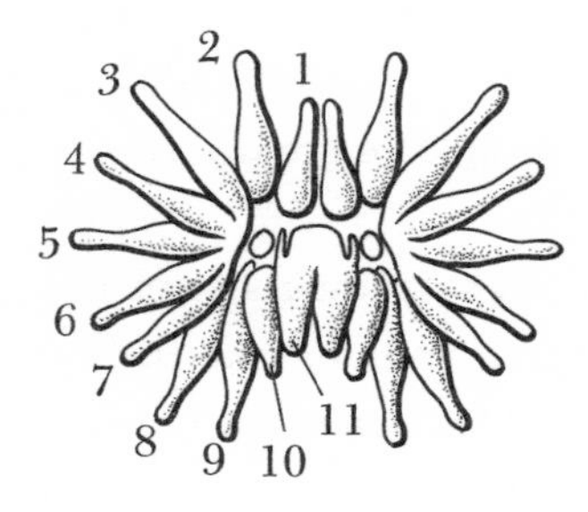

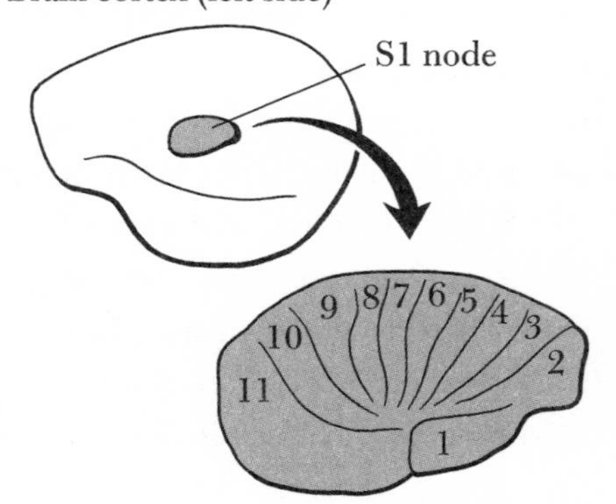

Figure 26 The brain of the star-nosed mole. The greatly modified nose contains 11 sensory appendages (fingers) on each side. The S1 part of the cortex, which receives nerves from these fingers, develops 11 corresponding regions in response to the incoming nerves.

dark. As might be expected, the corresponding cortex of the star-nosed mole has 11 barrels, each associated with a single tactile finger. Barrels are also found as a cortical representation of the forelimb in the eastern mole, of the hand of the owl monkey, and even of the bill of the platypus.[10]

How do the whiskers, the star organ, and the fingers of the monkey instruct the cortex to organize into barrels with nerves of the same topographic origin grouped together? Although more and more is known from the application to the problem of mouse genetics and pharmacology, the results are still ambiguous. The formation of the barrels can be divided into three periods: an early period when the axons arrive at the cortex; a middle period when the barrels form an accurate topographic pattern relative to the pattern in the thalamus; and a late period, ten to fifteen days after birth, when the nerves within the barrels make most of their synaptic connections.

The first period is largely governed by explicit patterning mechanisms in the brain itself, independent of the whisker pattern. The middle period involves reciprocal interactions between the cells in the cortex and the incoming axons that depend on electrical and chemical activity. It is this activity that is fundamentally exploratory and competitive and that results in partitioning of the initially uniform and indifferent cortical region into barrel domains. As in the simplification of the connections between single motor neurons and single muscle cells, organization is achieved by a form of variation and selection, with competition.

When the axons arrive at the cortex from the thalamus, they are actively firing in response to the movement of the whiskers relayed through the trigeminal nuclei and thalamus. As they fire, their nerve terminals secrete a neurotransmitter. If this transmission is disrupted in the cortex, then the barrelettes and the barreloids form normally but the barrels in the cortex fail to form. Although it is easy to imagine how electrical stimulation by the axons arriving from the thalamus might broadly activate the cortical region, it does not tell us how cells two steps removed from the whisker can end up grouped together to form a cohesive structure.[11]

Why would neurons in the cortex that respond to the same whisker be physically associated with one another? The Canadian psychologist D. O. Hebb (1904–1985) proposed that synchronicity of signals could strengthen synapses, as summarized in the aphorism, "Cells that fire together, wire together." When a whisker moves, all the nerves surrounding that whisker will fire, whereas nerves surrounding other whiskers will not be affected. The affected nerves will always (or nearly always) fire synchronously, and unaffected nerves fire together only occasionally by chance or never. If there is some way for all those cells firing synchronously to respond to one another, then they could group together. Whereas the Hebbian mechanism might strengthen synapses, it also changes cell shape or cell adhesion.[12]

In the barrel cortex, experience (whisker stimulation) acts locally to organize the indifferent cortical neurons into anatomical and functional units. It is an inevitable outcome of their physical association with a given whisker. The situation is highly unconstrained, since the cortex bears little preorganization that anticipates the organization of the whisker field. Hence, it can accommodate any whisker number and any strength of activity of the whisker neurons.

The same rules almost certainly apply to the formation of other highly organized structures like the star organ of the star-nosed mole and the primate hand. Despite the lack of clear anatomical manifestations, they must affect many, if not all, connections between incoming sensory signals and cortical neurons. In some species (hamsters, for instance) plasticity is maintained into adulthood. In these cases the power of exploration and reinforcement can correct damage, which is what apparently happens when cortical areas are partially reorganized in recovery from brain injury and stroke in humans.

The lessons from the barrel cortex suggest a solution to an important dilemma in evolution theory: that novel structures and processes should be difficult to evolve if they require several simultaneous modifications (in this case in the periphery and in three regions of the brain). Constructing a coordinated circuit from the periphery to the higher centers of the brain might seem implausible. Neural connections would have to be organized from whiskers through the trigeminal nuclei, to the thalamus, and then to a localized cortical region in the brain. At each stage a representation of the original whisker field would have to be created.

A detailed study of mechanism of the barrel cortex suggests a way around the dilemma. The whisker field itself can organize all the downstream processes because each participating neuronal area is widely receptive to an imposed organization. Such an adaptive process is based on total receptivity and selection, in much the same way as the microtubules respond globally to a stabilizing signal at the periphery.

Although the barrel cortex is an unusually vivid example, it seems likely that other wiring processes proceed by similar mechanisms and possess similar plasticity. The barrel field is not only adaptive physiologically (for example responding to the loss of a whisker) or developmentally (responding to errors in whisker number or placement), but also adaptive to evolutionary changes when whisker number and placement are altered genetically. The interchange of physiological adaptation and evolutionary adaptation may be particularly strong in the brain. It is known, for example, that in congenitally blind humans some of the visual cortex becomes responsive to tactile input, which may help the blind read in braille. The naturally blind mole rat has a related evolutionary adaptation: widespread usurpation for auditory stimuli of cortical regions that in other mammals are used for visual stimuli.

The Interchangeability of Physiology and Development

Embryonic development and adult physiology have usually been considered distinct. Yet in both the organism has several phenotypes, but only a single genome. Although embryonic development largely concerns anatomical change, as the simple egg is transformed into the complex adult body, physiological change occasionally also involves anatomical change, such as the growth of the uterus in mammalian pregnancy, tissue repair in wound healing, tail or limb regeneration in salamanders, or regeneration of the complete adult body by budding, as occurs in several organisms like hydra. It should therefore be no surprise that in some instances the same cellular mechanisms are used both for repair and regeneration and for embryonic development. Repair and development are not the same. Development usually occurs under carefully controlled conditions. The embryo, fully provisioned, may be protected in a shell or in the mother. In repair the starting point is variable; the animal is subject to the vagaries of the environment. Still, in both the goal is to achieve normal morphology and function.

Here we consider the formation of the vascular system, another wiring problem that has been studied in tissue repair and development, as well as in special cases such as pregnancy and tumor formation. The vascular system must be correlated with the growth of other tissues. Otherwise all of our favorite novelties from the trunk of elephants, to the wings of birds, to the tentacles of octopuses would be impossible. The growth of blood vessels addresses the same issue of how to avoid the need for multiple simultaneous changes to generate novelty in evolution; here the detailed mechanism is well understood.

The vascular system is a highly complex "organ" that permeates the entire vertebrate animal (some invertebrates do not have a closed circulatory system); there are typically sixty thousand miles of capillaries in a human being. Any cell of the body is always within about two cell diameters of a capillary. If one inspects the overall circulatory system of an individual, it looks highly organized, with a gradation in the caliber of vessels appropriate to the flow of blood. How is this closed circulatory system generated in the embryo? How does it regenerate in wound healing? And how does it enlarge in the placenta and uterus during pregnancy?

Elements of the growth of the vasculature in both normal and pathological conditions appear to be random. Even the arrangement of large vessels appears to be irregular. If you compare the veins on your two arms or two legs, they will not be mirror images, in contrast to your toes, joints, or large bones. Even the very large coronary arteries show variation from person to person. In some people the posterior interventricular branch is an outgrowth of the right coronary artery, and in others it arises as a branch of the left coronary artery. As much variation exists in identical twins as in unrelated individuals. Perhaps even on a gross level considerable variability exists in the anatomical relationship of blood vessels and overall anatomy. Yet variation in the large vessels is not unlimited; in particular, the main vessels have to connect properly to the heart and other organs.

In normal development, blood vessels are formed in two ways: from specialized embryonic stem cells called angioblasts, and from existing blood vessels. Early in development angioblasts coalesce and form hollow tubes, initiated by a hormone-like signal, vascular endothelial growth factor, which is secreted from the surrounding tissue. This step of blood vessel development is highly deterministic. Large vessels such as the aorta and the major embryonic veins are formed in this manner. Later the smaller vessels form by a different process, called angiogenesis. Vessels sprout from existing vessels, almost a type of vegetative growth like the sprouting of shoots of a plant. In response to a local vascular endothelial growth factor signal (as well as other signals), the vessel swells and becomes leaky. Individual capillary cells dissociate from one another, migrate toward the signal, and proliferate in response to it. Ultimately the capillaries reseal into a tube and rejoin in a continuous (closed) network. As more blood flows through the vessels, they increase in size. Although these features are well established, it is remarkable that local sprouting and growth can result in a system that efficiently provides oxygen and nutrients (and removes wastes and carbon dioxide).[13]

In this model of a totally responsive vascular network, ready to sprout new vessels in response to local signals, the position of each capillary must be determined by the prelocalization of billions of sites of vascular endothelial growth factor release. The "design problem" becomes one of placing those local signals. In fact, the growth factor is not prelocalized. All tissues have the capacity to signal. What is uniquely adaptive about the system is that signaling by tissues is directly related to their need for oxygen. Therefore, most of the vasculature is generated by a functional feedback process: the local need for oxygen drives a local cellular response, which leads to a local signal (vascular endothelial growth factor production), which in turn leads to increased capillary growth, increased delivery of oxygen to the previously oxygen-starved tissue, and finally termination of the signal and the process.

The intracellular circuitry that couples low oxygen levels to growth factor production is well understood. The most immediate response to increased oxygen demand in the whole animal, as mentioned in Chapter 3, is physiological, an increased rate of breathing. Low oxygen also causes production of diphosphoglycerate in red blood cells, causing hemoglobin to dump its oxygen more effectively in the tissues. At the same time, the levels of a secreted growth factor protein, called erythropoietin (the growth factor known as EPO, famous for its misuse

in blood doping for endurance sports like bicycling and cross-country skiing), increase several hundred fold, and this induces red blood cell formation in the bone marrow. These are systemic responses. However, there is a still slower and more local response, the induced formation of new blood vessels at low oxygen levels and inhibition of capillary growth at high levels. The responsiveness of capillary growth to hypoxia was discovered when premature infants were exposed to high levels of oxygen and found to have reduced capillary growth. When they were returned to normal oxygen conditions the previously oxygen-starved tissues grew catastrophically. The eye leaked and bled, leading to scarring and to blindness.

A system in which local low levels of oxygen cause the local synthesis and secretion of vascular endothelial growth factor (and several other molecules) ensures that throughout the body all tissues will achieve the proper blood supply. The existing vasculature is the permanent repository of responsive cells that can proliferate at any time. These cells migrate into the oxygen-deficient areas, attracted by signaling molecules. Migration is not random as is ant foraging—the capillary cells can sense the general location of the oxygen-deficient target—but the path is not deterministic either. It is a random walk biased roughly in the right direction by cells migrating preferentially toward the growth factor source.

The way the lack of oxygen stimulates the production of a specific protein is a biochemical Rube Goldberg mechanism. The transcription factor that regulates the synthesis of vascular endothelial growth factor is constantly made and destroyed. Part of its destruction requires an oxygen-dependent chemical modification of the protein. When oxygen levels are low, the transcription factor cannot be modified and hence is not destroyed. It accumulates and causes vascular endothelial growth factor to be synthesized, which then diffuses into the surrounding tissues, along with other factors, causing cells of the vessels to dissociate, proliferate, and migrate. The factors released by the hypoxic cells serve as cues to where the cells should migrate.[14]

New capillary formation can happen anywhere, at any time. Cells are set to respond to oxygen deprivation, and the existing capillary cells are set to respond to vascular endothelial growth factor. Tissue damage will generate signals that cause the normal vasculature to come to the rescue and satisfy the local need. This process occurs in the normal growth of the organism, but also in special cases such as wound healing and the growth of the uterus in pregnancy. Separate signals and special processes are not needed for embryonic development or for wound healing for any other special condition.

Tumors also need new blood vessels to grow and they too secrete vascular endothelial growth factor and related molecules. Blocking the growth of blood vessels toward the tumors is a promising approach to cancer therapy. Obviously vascular growth is adaptive evolutionarily, since the system will produce a blood supply to fit any oxygen need, without any requirement for central control or for modifying the core process.

The Likelihood of Novelty

That exploratory processes lower the hurdle for generating novelty is well illustrated by the evolution of the vertebrate limb. Limbs are diversified parts of vertebrate anatomy, encompassing fins, wings, legs, hands, paddles, and flippers, over an extraordinary range of sizes (Figure 27). We have said that generating a functional limb involves integrating the development of several anatomical and physiological systems: localizing the cells that form bone and cartilage, arranging the muscle cells relative to the bone and cartilage, directing the nerves so that they can innervate the proper muscles to move the limb, and emplacing a vascular system to oxygenate the muscles and other tissues. It is hard to imagine random genetic variation and selection to simultaneously relocate and reintegrate these independent complex

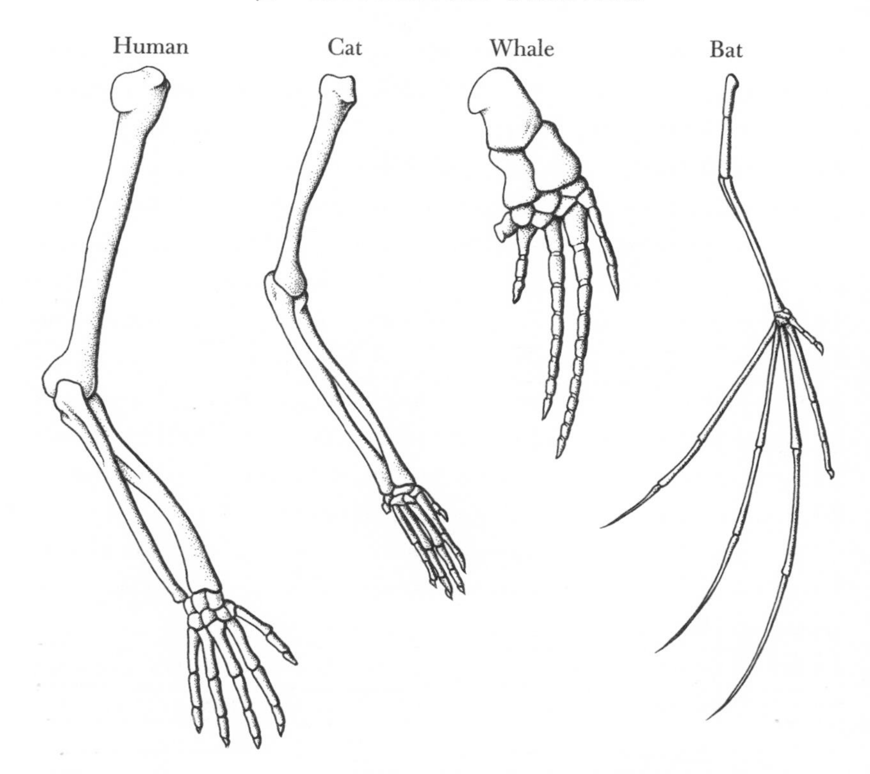

Figure 27 Variety in vertebrate limbs. All develop according to the same set of signals and cell interactions, forming the same initial set of bone rudiments. Differences arise from the subsequent independent growth of those rudiments, to give various longer or thicker bones.

systems. Thus, we return to the skepticism of William Paley and Darwin's struggles with "organs of extreme perfection."

What we have learned about the critical role of exploratory mechanisms in the development of the limb renders the problem much simpler and reduces the apparent requirement for simultaneous genetic changes. A succession of exploratory processes ensures that the supporting functions will always be available, even if the initiating skeletal changes of the limb are substantial. Over the course of evolution, the skeleton has been modified in a few basic ways related to the growth

and condensation of cartilage cells, resulting in the lengthening or shortening of bones, in the division of existing bones into smaller units, or in the fusion of bones into larger ones. Despite the changes in bone anatomy, muscle cells do not need to be modified at all by genetic change to participate in the development of the altered limb.

Muscle precursor cells are formed in the trunk in clusters close to the nerve cord. From this site, they migrate outward and follow an exploratory path into the neighboring appendage. There they associate with the bones and cartilage in whatever arrangement they find. The muscle precursors then proliferate and differentiate in response to local cues. Hence, migration of skeletal muscle precursors into the developing limb is an exploratory process, much like microtubule assembly. It can adapt to any of a wide range of limb bone anatomies. Experimental studies of grafting early embryonic precursors of the limb to unusual locations demonstrate that muscles find their way to associate and proliferate properly relative to the bone.

As described earlier, the nerve axons also follow an exploratory path into the developing limbs. Superfluous numbers of nerve cells are produced in the spinal cord, and their extended axons make redundant multiple contacts with muscle in the limbs. Electrical and secretory feedbacks stabilize the functional neuromuscular connections. Innervation of novel muscle sites in the limb can occur without the need to modify any aspect of axon extension or of responsiveness of nerve cells to survival signals. Meanwhile, in the brain, we can imagine a highly adaptable topographic map of inputs being made from sensory nerves that have also migrated into the modified limb. The map will show a representation of the new anatomy, accommodated to the modifications.

Finally, the vascular system sends migrating cells out to furnish vessels to any region that does not get enough oxygen. The hypoxic tissue will produce vascular endothelial growth factor and other signals, causing neighboring vessels to generate branches. The caliber of the vessels increases in response to blood flow, reflecting the size of the capillary bed, and in turn reflecting muscle activity and oxygen consumption. This kind of modification of the blood supply occurs physiologically, accompanying muscle growth (during exercise training, for instance).

Viewed in these terms, the normal development of limbs holds the key to the rapid evolution of new limbs. Normal development begins with the patterned deposition of cartilage-forming cells, the precursors of the bones. Then follows a series of highly adaptive processes that can generate the muscle-nerve-blood-vessel anatomy of the normal limb, but can also develop any of an unlimited number of related states, defined by the location of the bones. Thus, initially only the skeletal elements of the limb may respond to genetic change and the other tissues can adapt to them. We might imagine that subsequent refinement, involving genetic change under selective conditions, would improve the rough draft of the new limb, but what is most important is that innovation at this level would not be prevented by the difficult requirement for simultaneous innovation in multiple systems. Innovation, early on, can probably be substantial enough to reach the threshold of new function, hence to be selectable.

The history of limb evolution shows that new function relative to selective conditions can arise rapidly in evolution. Selection can be an effective tool in limb evolution, but only if significant changes are produced in each generation. Substantial change seems to imply simultaneous change in several systems, and simultaneous change implies an extreme rarity of occurrence. Exploratory processes provide an escape from this dilemma. These processes have an immense breadth of adaptation. Their adaptability is used in each organism in its normal development and in wound healing and regeneration. Thus, these broadly adaptive processes are under continuous selection for the function they serve and are available to support evolutionary change, when needed.

Evolutionary Change

Historically, some biologists have drawn the distinction between large evolutionary changes or macromutation and small changes or micromutation. One hundred years ago the topic was hotly debated, but the consensus gradually returned to Darwin's view that small incremental changes sum to larger changes; there was no need to postulate large steps in evolution. That view had three problems: If the steps were too small, it would take many of them to achieve a major innovation, and the question would naturally arise whether enough time had been available to generate all possible changes. Furthermore, if the steps were very small, there might not be enough incremental fitness at each stage for selection to act. Finally, the absence of intermediates in the fossil record provided weak evidence against the small-step view.

Large steps had their problems too. Only by taking small steps might the organism remain within its physiological range, a requirement to stay alive until new mutations or genetic reassortment stabilized the changes. For large steps, simultaneous changes might be required to realize the positive effects of novelty. We would expect multiple adaptive changes to be exceedingly rare.

Since heritable variation is necessarily limited by the physiological adaptability of the organism, it is necessary to know what limits the physiological range, particularly that range appropriate to the excursions in anatomy underlying evolutionary change. It is here that the highly conserved and pervasive physiologies based on exploratory principles have a special role. These processes are broadly receptive and therefore can respond to modifications never before experienced. In this way, they differ from many other homeostatic physiological processes like oxygen transport, which are two-state systems highly tuned along a certain range of response and evolved to operate within that range. The dynamic cytoskeleton, the connectivity of neurons in the brain, and the vascular system are not so limited. They operate on very different principles of generating random or nearly random variation (many states) and responding to local selection. The conservation of such mechanisms for almost a billion years is a direct result of the demands of the spatial complexity of multicellular organisms. Reuse of the same system in multiple contexts demands that the system be adaptable and receptive. Any system selected for the multiple states of trillions of cells is naturally prepared for new states in newly evolved

circumstances. There are simply not enough genes to have it any other way.

Exploratory mechanisms have a dual role in facilitating evolutionary change—which on the surface seems paradoxical. By being globally responsive and adaptive they blunt the effects of mutation and reduce its effect and lethality. In this way they make possible the persistence of novel changes by reducing collateral damage, thus increasing the amount of heritable variation. These are not incompatible because morbidity due to some types of changes is avoided, whereas other types of changes are preserved. In the case of the vertebrate limbs, the exploratory systems reduce the incidental stress on the vascular system, nervous system, and muscular system, but at the same time they allow the full expression of variation in the skeletal system. On the other hand, exploratory systems can be the targets of both environmental and genetic change. They too can form many morphologies, which require adaptation by other core processes. Their plasticity increases the scope of phenotypic variation.

SIX

Invisible Anatomy

If one looks at animals as one of the various life forms, and asks what is unique about them, it would have to be their large size and the varied anatomy by which their physiology and behavior are conveyed. It is not their chemistry or efficiency or resistance to harsh conditions, all of which are exceeded by bacteria, protists, fungi, and plants. Complex anatomy has not been achieved on the level of the single cell and hence emerged only with multicellularity, that is, in the last 600 million years. While relying on the conserved cellular processes brought forth from the earlier waves of innovation, anatomy also depends on the complex processes of embryonic development that evoke differentiated cells of several hundred kinds in a body of trillions of cells. So diverse and distinctive is anatomy that it has reliably served, beginning with Linnaeus, to classify living and fossil species, from which a pattern of evolutionary descent was later deduced. In general, the results of comparing DNA sequences of different animals have agreed with the anatomical phylogenies, although some significant discrepancies exist.

Generating the special anatomy of each animal species is embryonic development from a single-celled egg. In its early stages of development, an embryo produces conserved phylum-wide traits, such as the dorsal hollow nerve cord of vertebrates or the segmented body of arthropods. Later in development, the local and specialized bits of anatomy peculiar to the animal are added, and lastly the cell types differentiate. How is this exquisite detail of phenotype developed from a single-celled egg?

Development is a vast system of generative processes about which much has been learned. When a trait of anatomy changes in evolution, it is really the development of that trait that has changed. Anatomy itself is not inherited, but rather the means to generate the anatomy. The real target of heritable genetic change is the development by which the trait is produced. When a mutant is found with an altered trait, the role of the altered gene product is ultimately tracked down to an altered developmental process. Therefore, in seeking to explain anatomical change in evolution, biologists have come to understand that what they must explain is the changes in developmental processes.

In previous chapters we have examined conserved cellular processes mostly in adult differentiated tissues that are so adaptable and versatile, and so readily coupled by weak linkage, that they can produce almost any outcome in response to a wide range of environmental or mutational stimuli. In the realm of cellular anatomy, the dynamic and adaptive means for generating the cytoskeleton means that single-cell morphology is limited in its variety only by the spatial diversity of signals encountered by the cell. Also, the array of proteins produced in a cell in response to signals from other cells is versatile and alterable, given the nearly unlimited number of gene combinations that can be expressed through the flexible workings of the transcriptional machinery.

This adaptability that we have discussed for cells in the adult is true also of cells in the embryo. Embryonic cells respond in a plastic way, changing their response to external signals, altering their morphology, and revising the combination of genes they express. They also change the proteins they secrete, which serve as regulatory signals for other cells in the embryo. In this highly changeable environment, which is generally protected from the outside world, cells and tissues respond primarily to selective conditions imposed by other cells of the embryo.

A difficult problem in biology that was clarified in molecular terms at the end of the twentieth century is how the complexity of the adult arises from the single fertilized egg. Before that time, many believed that a cryptic prelocalized complexity in the fertilized egg foreshadowed the anatomical complexity of the adult. Eggs were thought to possess prepositioned packets of signals that would elicit the appropriate localized anatomical response, such as nerves in one place and muscle in another. However, attributing the anatomical complexity of the adult to localized signals in the egg hardly solves the problem. All embryos start from a single cell, and that cell would have to generate all the localized complexity itself, or else the parents would have to endow it with that complexity. Although biologists long ago imagined the egg to be unusual in having a complexity of organization equal to that of the adult, today we know that in most cases the egg is hardly more complex in its subcellular organization than a typical somatic cell.

How does a cell as simple in its organization as a typical somatic cell generate the complex anatomy of the embryo and adult? Since every cell has a complete genome, there must be signals that tell specific descendants of the egg which of the genes to express and where and when to express them. These cues do not come from outside the embryo, but instead from the embryo itself. With this independence from outside instructive signals, the process of development has been called a process of self-organization. Cells divide from the egg, generate signals, respond to signals, and consequently express a subset of the organism's genes and cellular behaviors (such as cell movement or cell proliferation). If the location and combinatorial expression of these processes can be changed in a facile manner, evolution too can proceed in a facile manner. Exploratory mechanisms and weak linkage lower the barrier for generating this kind of variation. But other processes, to be discussed, arose with multicellularity, specific to large-scale anatomical organization, and also contribute to evolutionary change.

The requirements of embryonic development are not just those of putting specific cell types in the right place; for example, of placing each of the three hundred differentiated cell types in each of the trillion positions in the body. Beyond this problem of differentiation, which in itself seems nearly insoluble, numerous experiments have demon-

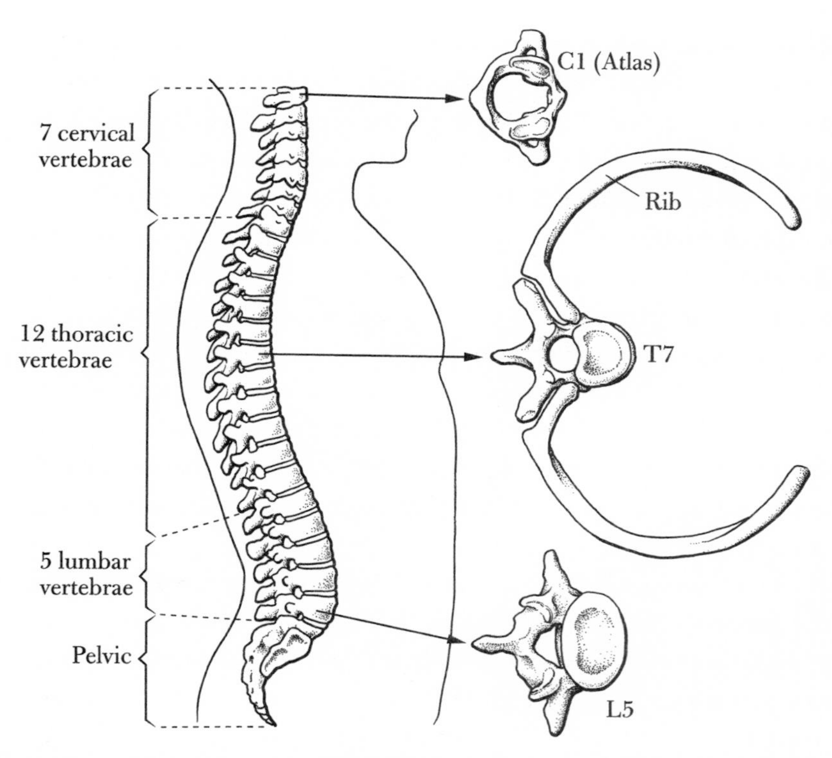

Figure 28 The diversity of human vertebrae. Although all vertebrae are formed by bone-depositing and bone-dissolving cells, the cells differ slightly in each part of the spine, leading to the different sizes and shapes of the 24 vertebrae (not counting the bones of the sacrum). The Hox genes are thought to control these differences. Enlargements shown here are vertebrae of the cervical, thoracic, and lumbar regions.

strated that different parts of the body have different subtypes of the major categories of cell type. For example, each of the 24 vertebrae of the backbone of the human is a bone dynamically deposited by and broken down by cells involved in forming and remodeling bone. However, vertebrae are not all alike, as shown in Figure 28. Those of the thorax have long ribs extending from them, whereas lumbar and cervical vertebrae do not. If thoracic cells are transplanted to a lumbar location before bone formation has started, they develop a ribbed vertebra, true to their old location, as if the cells of a specific locale become specialized at an early time in development for later making a vertebra of a particular shape. They are subtypes of the general cell types involved in bone formation, and their differences probably depend on their membership in a particular group of cells in a specific region of the body.

Many transient cell types exist in embryonic development—various multipotent responsive types, each competent to take any of several paths of development, depending on which local signal is met. Among these transient cell types are the various kinds of self-maintaining stem cells from which one or more differentiated cell types are generated. For example, in the bone marrow of adults are stem cells that proliferate steadily to replace themselves and to release daughter cells that differentiate to red blood cells, macrophages, platelets, and leukocytes, all nondividing differentiated cell types. As a cell type, the stem cell is different from any of its derivatives. When all these cell types are enumerated, there may be thousands or tens of thousands of kinds representing different stable expression states of the genome, called forth at different times and places in development.

A Map in the Embryo

We have a double problem in the development of complex animals: the anatomy of the embryo must be strewn with numerous changing signals that tell cells of that region what cell types to make. At the same time, we cannot expect so many different kinds of signals, given the relatively small number of genes and the simplicity of the egg.

The answer to this embryological difficulty turns out to be an answer also to the problem of morphological evolution in multicellular animals. Its solution emerged first in *Drosophila* from genetic studies that had their origins in the 1940s but came to fruition in the 1980s and thereafter. Undiscovered and apparently unanticipated until then was the existence of a coarse-grained map in all metazoan embryos. It was a map of cells, some of which produced localized signals for localized responses including cell differentiation and tissue anatomy. It was also a map of cell groups that were spatially differentiated in their response to those signals. It was a map of partially overlapping spatial domains in the embryo that effectively divided the embryo into different compartments, each distinguishable from the others by a few genes expressed within it. The map has no simple anatomical counterpart, and the borders of compartments often cross anatomical boundaries, just as political boundaries sometimes cross mountain ranges. It seems to us more or less like an arbitrary map, such as the ones left by departing colonial powers.

The compartment map is complicated in that some expressed genes are shared by two or more compartments. It is the exact subset of expressed genes that is unique to locale. If the spatial patterns of each locally expressed gene were colored differently on the surface embryo, a compartment would be a contiguous group of cells having a unique shade. Consider a hypothetical embryo expressing two localized genes, yellow and blue (and ignoring all genes common to all cells), with the anterior cells of the embryo expressing yellow and the posterior cells blue, but overlapping cells in the middle expressing both blue and yellow (green). The embryo would have three compartments, yellow, green, and blue. The *Drosophila* embryo might have one hundred such compartments at the five-thousand-cell stage; the vertebrate embryo, a few more, perhaps two hundred.

The existence of compartments immediately raises three questions: How are they emplaced? What is their role in the development that follows their emplacement? What effect do they have on the evolution of anatomical differences? Compartments are not present in the egg; as we said, the egg is anatomically simple and, by definition, different compartments comprise cells expressing different genes. Nevertheless, compartments arise by processes inherent in the egg operating under conditions set by the surrounding environment. It is a bootstrapping process where a few small initial differences are acted on to make further differences. When the particular compartment map is completed in the embryo, it is a map shared by all members of a phylum, the largest grouping of organisms based on anatomy and physiology. Though not strictly identical in each species of a phylum, the map is the most conserved feature of the phylum's anatomy.

Once organized in an advanced multicellular stage, but well before the cells have differentiated into their final cell types, the compartment plan gives each cell its address, its identity, and its location relative to cells in the rest of the body. This address will serve the cell and all its descendants into the adult. Each compartment will develop multiple tissues. Similar cell types in different compartments (with different addresses) may appear to be similar in their structure and activity, such as bone deposition or nerve impulse conduction, but will differ from one another in other ways, such as their capacity to proliferate, to migrate in the embryo, or to adhere to other cells. They can differ in subtle aspects of their behavior, as in the bone-forming cells that make thoracic and cervical vertebrae.

We call the compartment plan an "invisible anatomy" because the compartments are only identifiable if one can establish which genes are expressed there. At these early stages, compartments cannot be distinguished by anatomical features. The actual differentiation of the organism will depend not only on the compartments but also on the interactions of cells of one compartment with signals from other compartments. The compartment map is an extensible map: individual compartments can expand and shrink independently, while overall neighbor relations are retained. This flexibility occurs not only in development, when certain regions grow relative to others, but also in evolution where there is disproportionate growth—for example the neck of the giraffe relative to the neck of the whale.

The compartment boundaries, though curiously arbitrary with regard to the final anatomy, nevertheless divide the embryo into regions. In different regions the same target genes can be controlled differently. The result is a platform for local differentiation and for use by the genome in many different ways. Stated in terms of the conserved core processes, the compartment map makes possible the use of different combinations of processes at different places in the body. In fact, it provides those places. The map makes possible the use of different combinations of genes in different locations.

The concept of compartments has been the most valuable contribution of cellular and developmental biology to understanding evolution, at least the evolution of complex multicellular animals. The basic compartment structure independently defines the phylum and hence must have remained unchanged since the Cambrian, for more than 500 million years. Why the conservation? Are the conserved global features so fundamental that any change results in lethality? Do the unchanging features of the phylum-wide compartment plan impede evolutionary changes of anatomy, are they irrelevant to such changes, or have they actually facilitated them? It is the last of the three questions, we think, that deserves attention, having received little in the past.

Initially the concept of compartments seemed to many like a theoretical abstraction. Yet it is a concrete developmental mechanism, as concrete as the control of oxygen binding to hemoglobin. Like hemoglobin, compartments serve to make the animal more robust to the environment, and, like hemoglobin, compartments are a platform for simple modification in evolution. Unlike hemoglobin, we are in the realm of spatial control of embryonic development and not of quantitative adjustment along a simple reaction norm. Thus compartments are more complex than hemoglobin, though ultimately built on the same weak linkage and exploratory mechanisms. To understand compartments better in the context of other biological questions, we begin with a brief history of how compartments came to mean what they mean. We will then be in a position to appreciate the consequences of compartments for facilitating variation in evolution and to understand why anatomical deconstraint can emerge from such highly constrained processes.

The Discovery of Compartments

Compartments were a surprise of such unexpected generality that, in the space of ten years, the entire field of developmental biology was refounded on completely different principles. It helped that this period coincided with the development of recombinant DNA techniques and new procedures for visualizing the expression of genes at the individual cell level. Compartments emerged as an answer to questions raised classically by embryologists in the first half of the twentieth century. These questions revolved around how a single-celled egg with its single genome could give rise to the complexity of the adult.

Experiments on embryonic development began in the late 1800s. Dye marks were spotted onto eggs, such as those of sea urchins and frogs, at reproducible locations, and they were seen to end up at corresponding reproducible locations in the hatched animal. Therefore, a map could be drawn on the egg of the points of origin of the anatomy of the adult. The embryologist and philosopher Hans Driesch could say in 1894 that "a cell's fate depends on its position in the embryo," but it was completely unclear what position meant—position relative to what?[2]

Hans Spemann found ways in the decade 1910–1920 to transplant small groups of cells in amphibian embryos, moving them from one location to another. He did this at an early stage of development—when ten thousand cells were present, long before they had differentiated, and even before they had rearranged their locations (in the massive cell and tissue migration known as gastrulation). If left in place, these cells would reliably contribute to known anatomical structures such as the brain or the skin of the belly. To Spemann's surprise, the operated embryos developed normally, and the transplanted cells developed according to their *new* location, not their old. Cells that would have made belly skin now made brain, and vice versa. The cells somehow gained information about their new location, but how?

As discussed in Chapter 4, Spemann and Mangold in 1924 found a special group of cells, about 5 percent of the total that was the reference point, which they called the organizer. If cells of the organizer were transplanted to a new location, all surrounding cells developed to new fates according to their distance from that special transplanted group. The organizer was a source of signals that spread to other cells and affected their development. Even the low level of information carried by the signals was enough to pattern the entire embryo.[3]

Later analysis showed that each cell responding to the organizer was competent to differentiate in any of several ways, and that it made the decision according to the amount of signal, which was related to its distance from the organizer. The cells had at least three response options: no signal, low signal, or high signal. Since cells of a large region are initially the same in their repertoire of developmental possibilities, Spemann and Mangold could exchange them by transplantation. The cells would still give normal outcomes, as long as the exchange was done before the cells received and responded to the signals. Later, cells became committed to certain of their options and could not change. Here too, in embryonic development, physiological variation and selection were found: embryonic cells had a range of alternatives open to them (their variation), and the intercellular signals selected among the options. It was a form of exploratory system, with selection from a spatially localized source of signals.

Lewis Wolpert in 1969 summarized the state of understanding in terms of positional information (the kinds and quantities of signals spreading from a source) and interpretation (the response of individual cells according to the signal and its level). An embryonic cell at any particular moment has a set of developmental options, defined by its genotype and previous developmental history. Its choice is then dictated by the kinds and amounts of signals it receives from other cells, the signals depending on the cell's position in the embryo. Signals, however, provide only limited information, and the cell *interprets* the signal according to its preset options (a permissive induction).[4] Although the dialogue between cells during development was well appreciated, it was unclear until the 1990s what signals were emitted and received at different distances, and what the cell's response entailed.

The few early geneticists working with *Drosophila* mutants altered in their patterns of development started with similar questions but ended up with different conclusions. Curt Stern (a product of Morgan's *Drosophila* genetics group) found that a mutant cell surrounded by normal cells almost invariably gives rise to progeny that are mutant in their differentiation. This was a result opposite to Spemann's finding with cells moved to new locations in frog embryos.

Sydney Brenner, a molecular biologist famous for his work on messenger RNA, the genetic code, and embryonic development—and also famous for his irreverent wit—referred to frog cells as following the "American Plan," whereby you do not care who your ancestors are but you consult your neighbors about how to act. By contrast, the fly cells follow the "European Plan," whereby you do not care what your neighbors think but you only care what your ancestors thought. Stern concluded that his mutants must have controlled the cell's interpretation of signals at the site, but not the generation of signals there. Thus, combining the evidence from the frog and fly, we conclude that some genes in living organisms control diffusible signals, such as those produced by the organizer, and others must control processes that interpret those signals, such as those in the *Drosophila* experiments. Since the organism is spatially differentiated, each of these genes had to be active in the proper position on the embryo, and that would require global positioning mechanisms. Yet there was a disconnect in the two studies; Stern did not know how the particular responsiveness of cells was spatially allocated, and Spemann did not know how particular signals were allocated in specific regions of the embryo.[5]

Stern looked at a great variety of pattern mutants, and all followed the European Plan, whereby the response to position (their interpretation) was altered, not the generation of broad signals. This finding was a disappointment, since it gave no information about the signaling process and only indicated that many genes encoded products needed for individual cells to respond to signals. One mutant had an important difference, which later led to discovery of the animal's invisible anatomy—its array of compartments, which are domains of both interpretation and signaling.

The fruit fly is an excellent subject for morphological study, with five thousand bristles, each reproducibly positioned on the surface. Curt Stern's mutant (later shown to be caused by a defect in a single gene) had a number of definitive morphological defects and changes in the arrangement of the bristles, some giving an engrailed appearance to the thorax (*engrailed* is a heraldic term meaning indented edge). The mutant also had abnormal wings and legs in which the posterior parts looked rather like the anterior parts. Stern made small patches of mutant engrailed cells in the developing wing and leg. Remarkably, if the small patch arose in the anterior half of the differentiated leg, its bristle pattern was normal; if the patch was in the posterior half, it was abnormal, namely, a mirror image of the anterior bristle pattern at the corresponding location. The leg was composed of two halves, arranged in mirror image around an invisible boundary. Without the activity of the engrailed gene, posterior cells would be like anterior cells in their interpretation of signals. This discovery was one of the first indications that invisible large-scale patterning processes, involving specific genes, must underlie the visible detailed pattern.[6]

The Spanish developmental geneticist Antonio Garcia-Bellido chose the *Drosophila* larva rather than the adult, in hopes of understanding the earliest events of pattern formation. The larva has no wings but harbors two small bags of cells, called wing discs, that proliferate as the larva grows and then differentiate into the wings at metamorphosis, eventually projecting outward on the surface of the fly. He and others found that 15 embryonic cells are normally set aside in the embryo as the founders of each wing—they contribute to no other structure of the fly. In the newly hatched larva, they proliferate to one hundred thousand cells in about 13 rounds of division, and then differentiate.[7]

At these early stages of development Garcia-Bellido wanted to find out what each of the 15 cells in the wing discs knew about what role it was to play in the developed wing. Were the cells equivalent or different? When a cell divides, will its cellular offspring contribute in a significant way to a few parts of the wing or in a minor way to all parts of the wing? Do all cells proliferate equally or do some proliferate more than others? Are there early restrictions on what cells can do, and do their options for development become successively narrower with time? To which we add, Which of these properties might have changed as wings evolved? Would these changes have been difficult or easy to make?

Studying the wing and its early precursor, the wing disc, Garcia-Bellido marked cells at various times and followed their subsequent proliferation. He found that if he marked one of the 15 founder cells early, the eventual large patch of cells descended from it occupied any part of the posterior half of the wing or any part of the anterior half, but the cell descendants never crossed a midwing border to occupy parts of both halves. Thus, they were specified only as to the anterior or posterior compartment, not as to a position within the compartment. This mysterious border did not coincide with any anatomical boundary.

From these results, Garcia-Bellido concluded that anterior founder cells in the embryo constitute one "compartment" of wing development and the posterior founders constitute another. The gene that is most important in controlling the identity of the anterior and posterior compartments is the engrailed gene, which later molecular studies demonstrated was expressed only in the wing's posterior compartment. Thus, posterior cells in the 15 founders differed from the remaining anterior cells by the expression of the engrailed gene, a transcription factor. As growth and development proceeded, differences set in among cells of the compartment. They were no longer equivalent. A general conclusion from these studies is that development involves establishing smaller and different compartments nested within previous compartments, and then patterning those compartments.

Each of the 14 anatomical segments of *Drosophila* has an anterior and posterior compartment, distinguished initially by the expression of the engrailed gene in the posterior. What makes the segments different from one another? Why do some segments have wings and other legs and antennae?

The first hints came from Edward Lewis at the California Institute of Technology, who studied a *Drosophila* mutant called bithorax, shown in Figure 29. Bithorax was originally found in Morgan's laboratory in 1915. It was a strange fly with an extra pair of wings, just like the extra pair of wings that was induced by exposure to ether in Conrad Waddington's experiments. *Drosophila* normally has one pair of wings behind which is a pair of stumpy balancing organs, the halteres. In the mutant, the halteres were partially transformed to wings. In the extreme case, the fly had four wings.

The bithorax mutant is an example of homeosis, named by William Bateson, in which one part of an animal is missing and replaced by another part, such as a leg in place of an antenna, or a third leg for

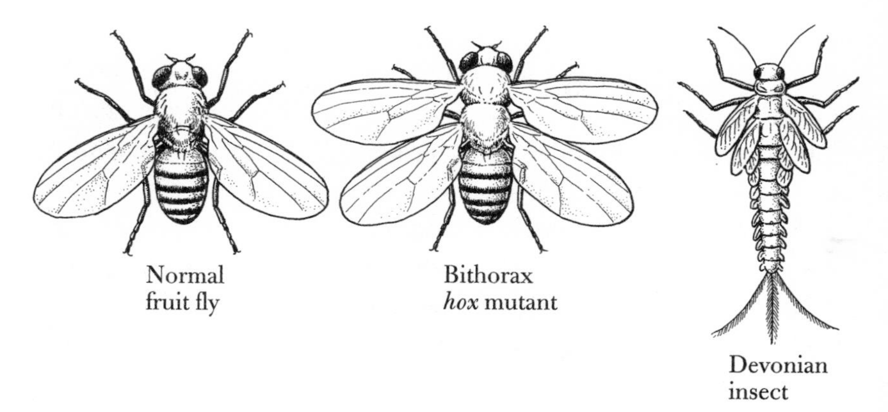

Figure 29 The evolution of wings in insects. *Left*, a normal fruit fly with one pair of wings. *Center*, bithorax, a mutant fruit fly defective in the *hox* 7 gene, with a second pair replacing the pair of small balancers. *Right*, a Devonian winged insect with many pairs of wings. In subsequent evolution, wing development was suppressed by Hox selector proteins on all segments of the fly except one.

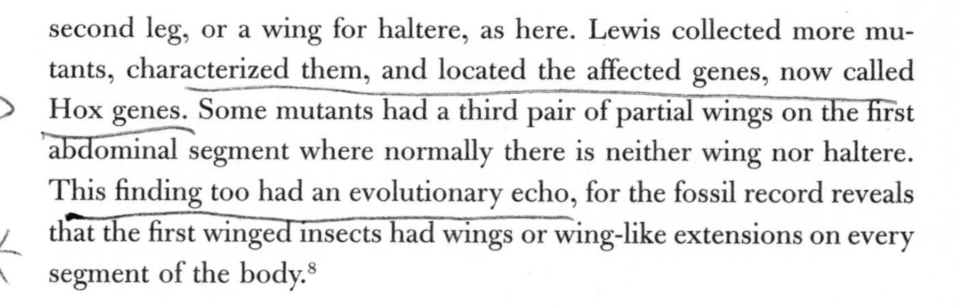

second leg, or a wing for haltere, as here. Lewis collected more mutants, characterized them, and located the affected genes, now called Hox genes. Some mutants had a third pair of partial wings on the first abdominal segment where normally there is neither wing nor haltere. This finding too had an evolutionary echo, for the fossil record reveals that the first winged insects had wings or wing-like extensions on every segment of the body.[8]

Selector Genes

Today it is known that Hox genes are present in all bilateral animals including humans, and that their role is much more general than wing formation. When a particular Hox gene is defective, the animal loses the ability to make a specific part of its body different from other parts, and one of the other parts is made instead. In time, 8 Hox genes were found in insects. They control the pattern differences in the fly's body from the posterior head almost to the end of the abdomen, the interval subdivided by the 14 body segments. The Hox genes are responsible for the differences of segments in terms of the kinds of appendages and bristle patterns formed in each. All 8 Hox genes have been removed in the flour beetle and the mutant still develops 14 segments, but all are the same—each resembles a head part with antenna. Thus, the Hox genes are needed to make segments different from one another and from the head.[9]

Garcia-Bellido argued that Hox genes are really "selector genes"; they affect a set of target genes and therefore select what kind of development will occur in the compartment of the embryo where they are expressed. The Hox genes make different regions of the fly different in their anterior-posterior dimension. Therefore, the Hox protein of the compartment selects what kind of subsequent development will occur there, leading to particular anatomical structures and cell types of that body region.

All the Hox genes are transcription factors that influence the combinations, amounts, and orders of the various conserved core processes at specific regions in the fly. Because of weak linkage, control by specific Hox genes can be easily imposed on many genes, bringing them under the local control of that compartment. The compartment array as a whole creates the places where these different combinations occur in parallel. The fly is divided into 8 or more large compartments, within which are contained the 14 smaller adjoining segmental compartments, each of which is divided into anterior and posterior segmental compartments. Engrailed is a selector gene that makes the posterior compartment different from the anterior. Thus, segments at the front end of the fly are like those in the back in that both contain the same kind of subcompartments (engrailed expressed in the posterior half), but they differ in the Hox selector genes expressed. The body plan is a complicated mixture of overlapping compartments, as illustrated in Figure 30.

Garcia-Bellido's concept of compartments was a breakthrough in development. Lewis' genetic work on the Hox genes then set the stage for molecular analysis of the Hox gene sequences in the 1980s and 1990s and for the discovery that they are a family of closely related

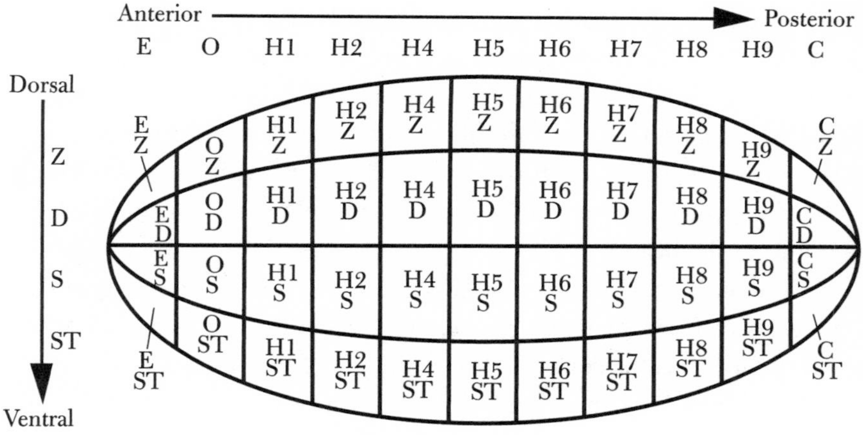

Figure 30 The idealized compartment map of *Drosophila*. During early development the embryonic body is divided into regions of expressed selector genes, some regions running vertically as shown (genes E, O, H1, etc.) and some horizontally (genes Z, D, S, ST). Vertical and horizontal overlaps of expressed selector genes demarcate the 44 unique compartments shown here.

transcription factors. They were later shown to be present in all animals, where they serve the similar function of dividing the embryo into compartments in a general anterior-posterior direction. Ed Lewis received the 1995 Nobel Prize in Physiology or Medicine with Eric Wieschaus and Christiane Nüsslein-Volhard, who had worked out by genetic means the early steps in the egg and embryo used to set up the compartments in the fly.[10]

To put these discoveries in a developmental context, we can say that early development is a series of events by which certain selector genes are turned on in certain regions of the multicellular embryo. All cells in a compartment are initially identical in their responsiveness to signals, since they all express the same selector gene. While continuing that expression, they become different when some cells of the compartment receive different amounts of signal from nearby compartments and respond to them. Often the signals are produced at the compartment boundaries and diffuse into the adjacent compartments. Cells close to the boundary receive more of the signal and take on paths of

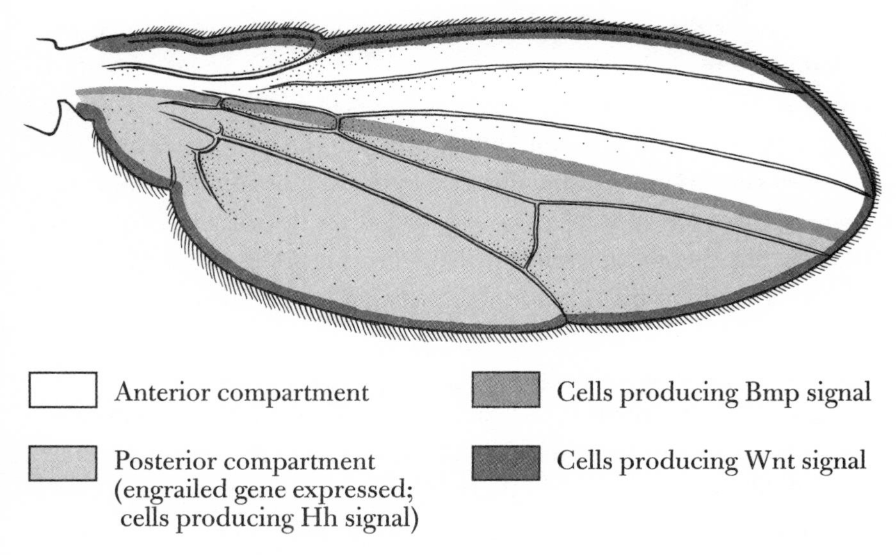

Figure 31 Compartments of the insect wing. Cells of the posterior but not anterior compartment express the engrailed selector gene and secrete the Hh signal protein. At the compartment border, the Bmp signal protein is produced and secreted. At the edge of the wing, the Wnt signal protein is produced and secreted. The pattern of bristles, hairs, and sensory nerves that develops on the wing depends on these compartments and signals.

development different from those more distant (hence less exposed). The more-exposed and less-exposed cells as groups become subcompartments with additional and different selector genes activated in the groups.

As illustrated in Figure 31, in the wing the posterior compartment under control of the engrailed protein makes a signaling protein that diffuses slightly into the anterior compartment. A few anterior compartment cells in a thin line close to the boundary respond to that signal by producing a second secreted protein that diffuses into both compartments in a mirror-image gradient. Thus, several gradients of signals are available for patterning the developing wing. Although we know the signals and their distributions for only a handful of examples of compartments, most or all are expected to have such signal sources at or near their borders.

Selector genes must remain on for a long time to influence the eventual differentiation of the cells of that region as adulthood is reached (that is, as the larva metamorphoses into a fly). As the embryo develops, various tissues will differentiate but the cells that form these tissues will always be distinguishable by the expression of the selector gene. In this way, the cells forming the haltere or balancing organ are similar but slightly different (in the degree of proliferation and morphogenesis) from the cells forming the wing in the next-most-anterior compartment.

Many processes are affected by selector genes, including cell division, cell differentiation, cell signaling, cell adhesion, and other developmental activities. In some cases, selector genes modify only slightly the activity of a cell, perhaps affecting its size or growth rate. In other cases, they control the cell's fate by preventing or allowing a certain path of differentiation. In yet other cases, they turn a cell into a signaling center that affects the fate or activity of surrounding cells. Circuits of selector genes are themselves conserved core processes that entrain other conserved core processes using weak linkage. Failure of a selector gene removes that compartment from the map. Cells of that location receive a different identity, usually that of an expanded nearby compartment. Then, for example, halteres turn into wings, or posterior segmental compartments into anterior ones. A compartment map exists in humans, and most certainly all bilateral animals, utilizing similar genes and proteins.

By now at least 30 compartments, with other collections of selector genes, have been found in the *Drosophila* body in addition to those expressing the 8 Hox genes or the engrailed gene (14 times repeated). When the selector transcription factors are visualized using special stains that detect the RNA transcripts of the selector genes, the embryo looks like a grid or map of domains—some overlapping, some not, some back to front in orientation, and some anterior to posterior. The map has been called the organism's second anatomy. It is the real body plan of the animal, a deep spatial organization of the body, invisible unless revealed by special stains.[11]

The Surprising Conservation of Compartments

What does the compartment concept tell us about the complex spatial organization of animals and about how easy or difficult it might have been to change that organization during evolution? We need to conduct a comparative study of compartments and how they have been used.

As research on *Drosophila* compartments and selector genes advanced rapidly in the 1980s, researchers turned to vertebrates such as mice, chicks, frogs, and fish to see if they had a similar compartment scheme for dividing the embryo into different regions from anterior to posterior. At first, scientists were skeptical; the Hox domains might only apply to arthropods, which have a segmented body rather unlike that of vertebrates.

By 1984, Hox genes were found in the frog, mouse, and other vertebrates. Insects have one cluster of 8 kinds of Hox selector genes. This single cluster must have expanded to 13 and duplicated twice in the evolutionary path leading to vertebrates, because vertebrates have four clusters of 13 kinds of genes. The total, however, is 39 and not 52; presumably some of the duplicated genes were subsequently lost. The gene sequences are similar to those of *Drosophila* in the regions that encode the DNA binding sites of the proteins. In mice, the Hox genes are expressed in domains from anterior to posterior, hindbrain to tail, in the same order as the genes cluster on the chromosomes in mice and also in *Drosophila*.

The targets of the Hox selector proteins are mostly different, however, as is evident from a comparison of the anatomy and physiology of the fly body and the mouse body. By now this map of Hox compartments has been revealed for many members of the chordate phylum. To return to the example of the different vertebrae in the backbone of the mouse, some with ribs and some without, the differences are due to the fact that bone-forming cells of different anterior-posterior body levels express different Hox genes, and these affect the details of their bone-forming operations.[12]

As other selector genes were found in *Drosophila*, they too were sought in chordates. It came as a surprise (if not a shock) in 1993 to find that selector genes expressed in the anterior part of the *Drosophila* head, which is a body region anterior to the Hox compartments, are also expressed in the mammalian forebrain and midbrain, some in association with the eyes in both cases. Until that time, it was widely thought that the vertebrate head is entirely novel, the invention of our phylum. With the unmistakable similarity between *Drosophila* and mouse selector genes in the head, it seems that this part of the body, as a realm of the compartment map, was already present in the Precambrian ancestor of arthropods and chordates. The heads are quite different in their anatomy and neural organization, but for arthropods and chordates the selector genes of both compartments constitute a common platform for head development.[13]

The existence of a conserved domain map of selector proteins that regulate the patterning of tissues in the adult organism allows us for the first time to infer the organization of the ancestor of chordates and arthropods, which would be the Precambrian ancestor of all bilateral animals. This organism has never been seen in the fossil record. At most we may have traces of the burrows it made in the mud about 600 million years ago. Yet we know from the similarities of *Drosophila* and chordates that this common ancestor must have possessed the compartments and other anatomical features that have been conserved and brought forward in both lineages. Indeed, arthropods and chordates were already well diverged when they first appeared 530 million years ago in the Cambrian.[14]

We noted in Chapter 2 that various molecular features suggest that the common ancestor was an elongated, bilateral, worm-like animal, probably with a through-gut (mouth to anus) and three tissue layers, depicted in Figure 32. We can visualize this hypothetical ancestor of all bilateral animals based on strong inferences from the gene circuits common to arthropods and vertebrates.

For example, from the striking conservation of Hox compartments along its middle and posterior body, we know the ancestor was an elongated animal that had distinguishable domains in the anterior-posterior dimension. From other common selector genes, we know it

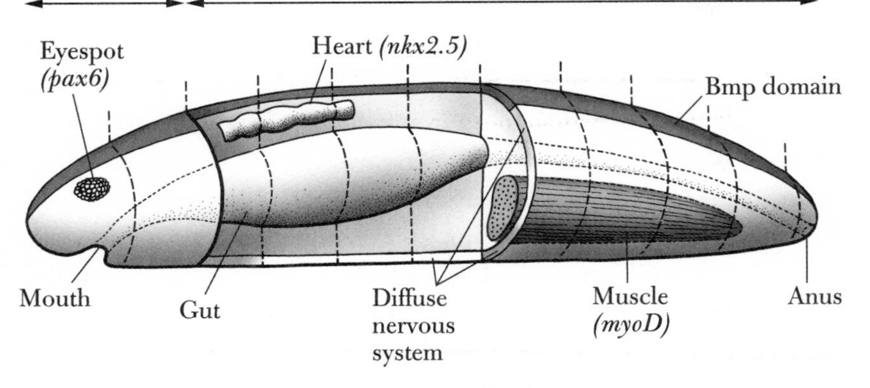

Figure 32 Our Precambrian ancestor. This worm-like animal has been deduced from the characteristics shared by all modern bilateral animals. The various selector genes (*emx, otx, hox, pax* 6, *nkx* 2.5) encode transcription factors expressed in different domains. *MyoD* is the master regulatory gene of muscle. The ancestor has not yet been found in fossils. Its length would have been about 0.25 inch (0.5 cm), as estimated from the dimensions of fossilized channels, possibly burrows of worm-like animals.

had a variety of other compartments in its anterior part, including those anterior to the compartments expressing specific Hox genes. It had organization from back to front based on particular signals secreted from the opposite midlines, signals still used in *Drosophila* and chordates. Whether the nervous system was diffuse or centralized is not yet clear. Even though the detailed structural organization of our brain is very different from that of a fly, it is based on the same underlying basic compartment plan, which has been conserved for over a half billion years. The ancestor may have had a light-receptive eyespot in which a conserved eye-forming gene was expressed, and a heart-like contractile vessel in which a conserved heart-forming gene was expressed. It probably did not have body extensions such as appendages or limbs (these came later), and it may or may not have had body segments. A significant implication of the comparison is that compartments are much more stable across evolutionary history than are the anatomical structures developed upon them. Compartments, it seems, are unconstrained in the anatomical structures and differentiated cell types they can support.

Since about 30 different bilateral phyla are alive today, we expect 30 different compartmental body plans, all modified in different ways from the plan of the common ancestor. Only a few plans have been characterized so far. Since all other phyla are descended from the common ancestor that gave rise to chordates and arthropods, we would expect all to have the central Hox genes in an order from head to tail, as well as several anterior selector genes because, according to the *Drosophila*-vertebrate extrapolation, the Precambrian ancestor had them. Each phylum must then have added new compartments in new arrangements and lost a few compartments, giving each phylum a distinctive map.

Such a view of the developmental processes that underlie anatomy was never before possible from study of the fossil record or the anatomy of extant forms. The molecular information is so precise and detailed that the interpolations are nearly unassailable. Between insects and vertebrates the sequence of Hox genes is conserved, the order of Hox genes on the chromosome is conserved, and correspondence of the chromosome order to the anatomical order is conserved. These similarities cannot be accidental or convergent from separate starting points. In combination with an increasingly detailed fossil record that shows maintenance of the body plans of almost all phyla for the last 535 million years, the commonality of the basic anterior-posterior patterning is proven far beyond any demonstration possible from fossils and comparative anatomy alone.

At the same time, anatomical inventions often can be traced to their developmental roots. For example, the chordate tail, a uniquely chordate structure extending beyond the anus, has three Hox genes not expressed in *Drosophila.* A phylum closely related to chordates, the hemichordates (acorn worms), also has these genes. Though it does not have a typical chordate tail, it does have a transient extension in the juvenile. The new Hox genes must have arisen by duplication of ancestral Hox genes after the split between the chordate and arthropod lineages; they must have been employed in posterior specification. Later the chordate tail was constructed on that regulatory scaffold. Such strong conclusions are possible because there is no doubt that these genes, possessing such strong sequence similarity, must have descended from a common ancestor.

The Role of Compartments in Evolution

Compartments appear in the embryo only at a middle stage of development. They are not present in the egg, not even when the egg has divided into several thousand cells. The early development of the cleaving egg is directed toward the overall contours of the embryo and spreading the compartment map on it. The middle stage of development when the compartment map is first present is called the phylotypic stage. It is when embryos of all the different classes of a particular phylum of animals look most alike.

At the phylotypic stage of chordates, the embryos of humans, fish, birds, frogs, reptiles, and even sea squirts look remarkably similar, though of course not identical. A set of compartments subdivides the nervous system and the adjoining muscle blocks. The chordate phylotypic stage also has the dorsal hollow nerve cord, gill slits, the beginnings of the tail, and the notochord. Phylotypic stages of different chordates are similar in size, about 0.04 inch (1 mm) in length, even those of the whale and the mouse, which as adults will differ a millionfold in weight. They have achieved their overall organization and compartment body plan, but have not yet used it to develop the diversity of anatomical structures that make the classes and orders of the phylum look different (no appendages yet, for example). After that middle stage, late development adds all sorts of embellishments to the map.

Our guess is that the Precambrian chordate ancestor did not make many additions to the conserved compartment body plan. Its embryonic development may have ended with an organism that looked rather like a phylotypic stage of today's chordates, onto which were added various differentiated cell types, such as muscle, nerve, and epithelium.

In the 500 million years since the Cambrian, embellishments were superimposed on the map. The map has served as a scaffold or platform for building more complex structures, such as brains, bones, limbs, and appendages. Such a succession is not unlike a Gothic cathedral built on the foundation of a Romanesque church. In addition, the relationship among phylum-specific compartment maps (those of chordates and arthropods, for example) gives us hints about the early emergence of the various phyla from the bilateral common ancestor, an organism still not found in the fossil record.

What is the consequence of the conserved compartment map of the body plan for the animal's capacity to generate variation? Has this map limited variation, or has it facilitated it? The grouping of organisms by similarity was an early means of classification. It was Darwin who turned this classification into a reflection of past diversification of descendants from a common origin. Now that we see an underlying similarity in the map of selector genes, what does that tell us about how variation occurred and, in particular, what was diversified in evolution? In the past, a phylum was defined as a group of animals sharing a body plan, which in turn was defined as a unique suite of anatomical traits. For this anatomical body plan, it would be circular to say it is conserved across the phylum. Yet the existence of a highly conserved map of selector genes and compartments provides independent confirmation of the significance of these phylum-wide anatomical similarities.

After the compartment concept was established, inquiry shifted from the evolution of anatomical traits to the evolution of the processes for generating those traits. Since 2000 we have been investigating the little-known phylum of hemichordates. These worm-like animals have no brain or central nervous system but do have gill slits. The tenuous argument that they have a subtle affinity to our own phylum was based on anatomical investigations in the 1880s by William Bateson and T. H. Morgan. When, 125 years later, we examined 22 selector gene compartments in a hemichordate, including the Hox genes, we found that they were in the same order and pattern as in chordates. The anterior compartments of the hemichordate's unique muscular proboscis had a compartment map similar to that of the chordate forebrain and midbrain. The gill slits had around them the same compartments as around chordate gill slits. We can see in Figure 33 that the compartment body plan of hemichordates is similar to that of chordates in the anterior-posterior dimension, even though the overt anatomy is quite different. Both show strong but reduced similarity to arthropods.

Finding 22 compartments laid out in this bizarre worm-like animal exactly as they are in human beings was like discovering a secret architectural plan hidden for over 500 million years! A large variety of tissues, organs, and cell types has been added to the body plan, but underneath the anatomical diversification the body plan has been conserved among all members of a phylum and is largely shared among related phyla.[15]

Decorations on the Tree

Why has the compartment map been conserved for such long periods in the face of large anatomical changes? To understand the relation of diversification of anatomy to conservation of the map, we must first consider development in the postphylotypic stage, that is, after the compartments have been established. This period is when various tissues, organs, and cell types are developed, in parallel in the different compartments of the body—those used by biologists to distinguish classes, orders, and families of animals within a phylum.

The test of whether the conserved compartment map provides more constraint or deconstraint comes from the success or failure of building diverse structures upon it. In examining the record of their use, we are forced to conclude that compartments of the body plan are very flexible. Basically, any kind of development can potentially occur within each compartment, without constraint. Target genes can be independently used in each compartment, because they can be regulated differently under the influence of the selector genes spe-

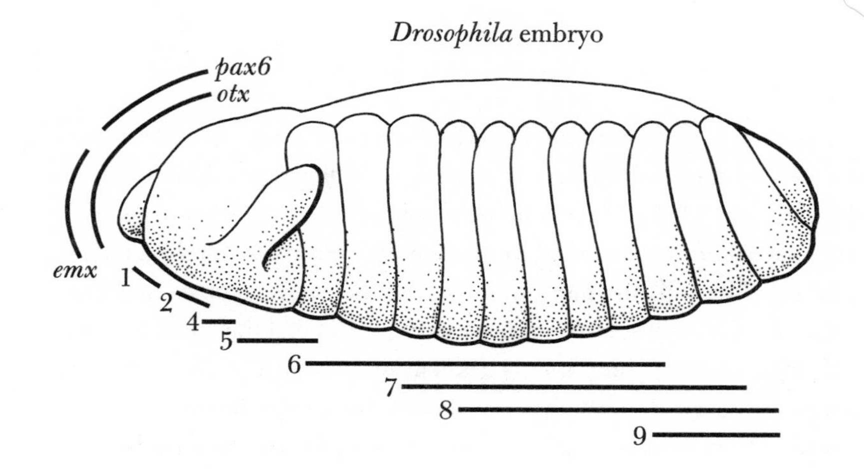
Drosophila embryo
pax6
otx
emx
1
2
4
5
6
7
8
9

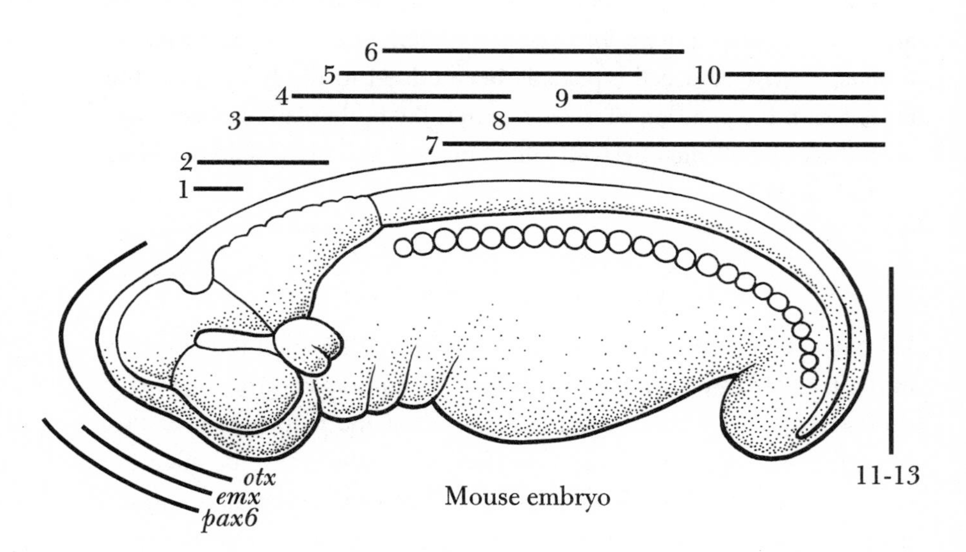
6
5
10
4
9
3
8
7
2
1
11-13
otx
emx
pax6
Mouse embryo

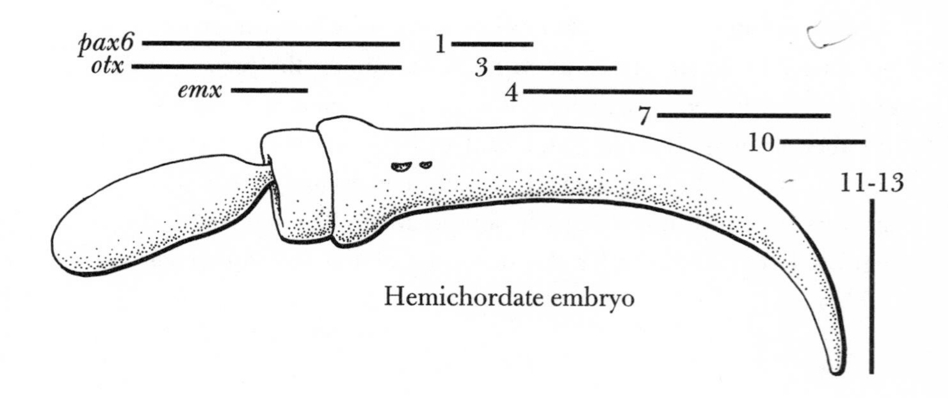
pax6
otx
emx
1
3
4
7
10
11-13
Hemichordate embryo

cific to that compartment. The same conserved core processes are used, but in different combinations, amounts, and times in the different compartments. During subsequent development the compartments can increase or decrease in size, largely independently. This is a critical feature of the many compartments of the body plan: *they can accommodate a great range of future anatomies developed in parallel.*

Compartmentation is a form of modularity, which is a common strategy in many designs. By subdividing the animal into smaller, largely independent domains, the evolution of structures in that domain can be uncoupled from the evolution of structures in other domains. (This is what happens when railroad cars are specialized for different functions: cars for carrying grain, oil, packages, people, and trailers. They all become very different, even though they remain compatibly coupled and still function on the same track.) Segregation and specialization reduce the so-called pleiotropy problem, that is, the problem of a mutation's having conflicting effects in different regions of the embryo, where a positive change in one place might provoke a negative change in another. If a change is lethal, a favorable change elsewhere can do nothing to overcome it.

Compartments mitigate these effects. If bones in the leg are to grow longer than bones in the arm, the expression of genes involved with cartilage and bone in the leg must be endowed with different properties. The arms and the legs are in different compartments. The organism can avoid conflicts over the use of common genes in limb development by using different local selector genes to regulate the genes expressed in the arm and the leg. New transcriptional control is readily available because selector proteins, as transcription factors, can in principle entrain any target gene. We know that in eukaryotic tran-

Figure 33 Conservation of compartments across phyla. Each phylum of animals has a unique compartment map of expression domains of selector genes (*pax* 6, *emx, otx, Hox* 1–13). As seen here, maps of different phyla have some similarities and some differences. Hemichordates are worm-like marine animals with gill slits, sharing an ancient ancestor with chordates.

scription it only takes a small change in the regulatory sequence of a gene to establish a new site for the selector protein to bind.

Even minor modifications within compartments can have major phenotypic effects if directed to specific anatomical regions. For example, all true flies (Dipterans) have wings arising from the second thoracic segment and not the third. By contrast, dragonflies (Odonata) have four wings, arising from both. The fossil evidence suggests that early insects had wings on all thoracic and abdominal segments. These evolutionary modifications reducing the number of wings to four and then two were clearly achieved by inhibiting wing development in all compartments except one or two. The Hox selector genes imposed the inhibition. The difference of the wing from the haltere is attributable to the Hox selector protein of the compartment in which the haltere is located, diminishing and modifying wing development at that location to form a haltere, but not eliminating wing development entirely. In a rare parasitic group called the Strepsiptera, males have two wings arising from the third thoracic segment but not the second, an opposite case of wing reduction.

The ancestors of the insects probably had legs on most or all segments, whereas true insects have but three pairs of legs on the segments of the thorax. Leg development was blocked in the other segments by Hox transcription factors acting repressively in the various compartments except those of the thorax.[16]

We should view the compartment body plan as maintained in evolution by selection, because its collection of selector genes divides the embryo in a complex but stable map. At the same time, few if any limitations are imposed on the kinds of processes that occur within each compartment, allowing them to operate in parallel. We believe that deconstraint in the changing of developmental paths within compartments exists even though the selector genes and maintenance circuits within compartments do not change, and may have their own internal constraints on change.

To summarize an unusual and critical part of our argument: The compartment map as such is conserved (and is constrained from change) because it facilitates (deconstrains) changes in the developmental processes that occur within the compartments. By separating regulatory functions, the compartments can change composition independently. Thus, the compartment is a flexible device for regionalizing later developmental processes and for reducing interference that would arise if each gene had to change in exactly the same way in all compartments (pleiotropy).

Of course, the target gene function would be nominally the same. For that to be individualized by compartment would require modifying the gene in a compartment-specific way, expressing the gene with another gene that would change its function, or making new compartment-specific genes by splicing or by other means. Compartments allow further selection for regulatory or structural diversification of genes.

Exploratory Processes and Compartments

An excellent example of the use of compartments after the phylotypic stage as a platform for further anatomical specializations is the neural crest derivatives. These embryonic cell populations are responsible for forming the entire peripheral nervous system, much of the skull, and numerous other tissues of vertebrates. The diversity of their specializations is exemplified by the numerous congenital effects of neural crest dysfunction, including cleft palate, neuroblastomas, and congenital heart defects. Neural crest cells are a hallmark of vertebrates; no other animal group has them. Some of the more unusual derivatives of neural crest cells are shown in Figure 34.

The development of the neural crest is a powerful combination of exploratory cell behavior and the compartment map. Explosive diversification can occur when there is a synergism between the independent and diversified compartments and a multipotential cell population that migrates over them and explores their diversity. The exploratory nature of the neural crest involves physiological variation and selection. The wide responsiveness of neural crest cells to signals and their capacity to undertake any of numerous paths of development in response to signals constitute their form of variation. The signals they find in the diverse compartments constitute their selection.

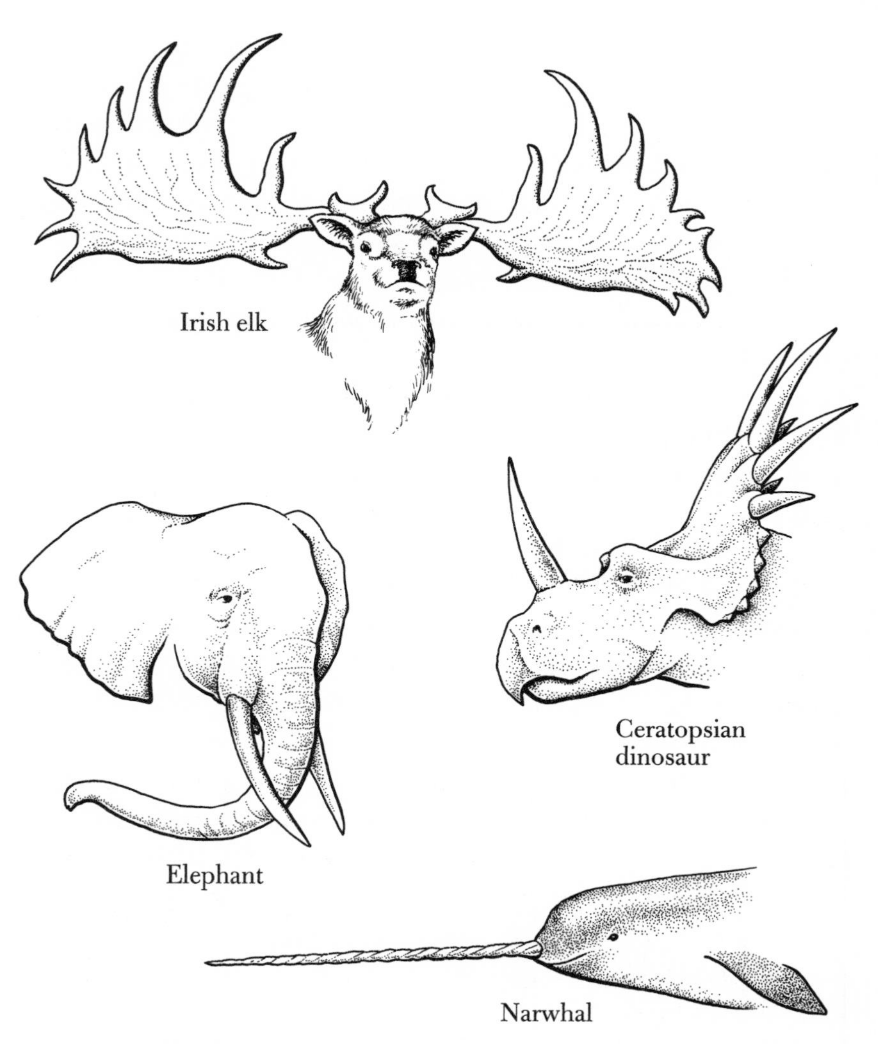

Figure 34 Peculiar differentiations of the neural crest. Several examples of head elaborations are shown. Only vertebrates possess neural crest cells. These cells are migratory, proliferative stem cells capable of many kinds and sizes of differentiation in response to the signals they receive.

Neural crest cells form at the edges of the neural plate (the precursor of the nervous system) on the backside of the embryo just after the compartment body plan is in place at the phylotypic stage. They migrate toward the belly, settling at sites in various compartments. Like other exploratory systems, they have alternative responses depending on the cellular environment. Unlike microtubules and ants, which have only two responses, neural crest cells have several. They can differentiate into neurons of various sorts, pigment cells, endocrine cells, connective tissue, cartilage, bone, and smooth muscle. They also form some specialized tissues like the dentine of the teeth, the valves of the heart, and the antlers of deer. It is a developmental repertoire that is unusually large for embryonic cells—basically that of a multipotent stem cell. Neural crest cells seem to launch an entire second round of development in vertebrates, following the earlier development of the compartment body plan. Of course, they depend entirely on the compartmentalized body plan for the location of potential sites of settlement, and then for the selection of particular developmental responses from their large repertoire.

A highly exploratory population like the neural crest offers two theaters for evolutionary innovation. The first is through the diversity of the neural crest cells themselves, as they arise in specific compartments in the neural tube. The second is through the various compartments they explore in the periphery during their migration. The product of these two forms of diversity can be exceedingly great and may explain the importance of these cells in vertebrate development. The neural crest has played a major role in the evolutionary modifications of cartilage, from the first gill arch of an ancestral vertebrate jawless fish into the jawbone of the jawed fish, which is a modified gill arch. It has also figured in the increasing skull size associated with brain enlargement. The rapid remodeling of the beak of the Galápagos finches, which Darwin found on his voyages of the *Beagle*, depends on the adaptability of neural crest cells, for they develop the beak.

The local adaptability of the developing crest cells to their surroundings reduces the potential lethality of hybrids or new varieties. Dog breeds provide examples of this robustness. The mating of an

English bulldog and a basset hound, which have such different facial structures (almost no snout versus a very long one, owing to differing neural crest development) produces a mongrel with mixed features. Nevertheless, the head is intact and functioning, at least under the supportive conditions offered by humans.

Deconstraining Early Development

Until now, we have considered deconstraint to be a property of the compartment plan such that it reduces limitations on possible kinds of development *following* the phylotypic stage. But the compartment plan would seem to be a major constraint on the egg and stages *preceding* the phylotypic stage. Activating a conserved compartment plan in its precise spatial arrangement at the phylotypic stage should require a suite of highly conserved developmental processes and signals, which should be very constrained. In particular, the array of compartments at the phylotypic stage is complex (up to two hundred kinds of compartments are eventually developed in the correct places in the vertebrate embryo), so one might think that two hundred well-articulated, independent lines of early development from the egg are needed to accomplish this complexity of organization. How would those two hundred lines be placed correctly in the egg and run in parallel without interference? Presumably this placement would require a set of agents arranged in as complex and as accurate a map as the final compartment plan itself. Furthermore, if such initial complexity were necessary, would the egg not be extremely constrained and conserved in its organization?

Two facts contradict such an expectation. First, many kinds of eggs can be grossly manipulated, and they still develop into well-proportioned individuals. For example, they can be cut in halves or quarters or several combined into one. Second, among chordates we can find a wide diversity of eggs, all of which develop successfully to a phylotypic stage with strikingly similar compartment maps. In birds, the egg contains a large yolk mass, so large that embryonic cells are divided from a small cytoplasmic island adjoining the yolk mass. In other organisms, such as the frog, the planes of division go though the entire egg. In some reptiles and in almost all mammals, the embryo develops extensive extraembryonic tissues, forming the placenta in addition to the embryo itself. Even within mammals, egg physiology differs. The duck-billed platypus lays a yolky egg. The kangaroo also has a yolky egg, but the embryo hatches inside the mother early and undergoes brief placental development. The "eutherian" mammals, such as humans, have a yolk-free egg and extensive placental development.

In addition to their remarkable properties of deconstraining late embryonic development, we argue paradoxically that the conserved compartments deconstrain the egg and early embryo as well, even though they are not yet present. The compartment map of the phylotypic stage may be special in requiring rather little information for its development. Only a few signals may be needed to orient and scale several of the compartments. The remainder of the complex array may organize itself from these minimal inputs, just as correct placement of a few key pieces of a jigsaw puzzle enables rapid placement of the pieces around them. If true, the early development would only be needed to get things started and would not have to be nearly as complex as the compartment map itself. The simpler the early development, the less constrained the egg. The egg could then undergo extensive evolutionary changes in organization of its own toward other ends, such as providing nutrients or protection, without jeopardizing the development of the compartment map that follows.

At the same time, to accomplish so much self-development, the compartments would need their own complex circuits and interactions. Complex compartment circuitry would mean high constraint on their evolutionary change. They would be unchanging, as indeed they are. The trade-off would be lessened demands on early development and lessened demand on the egg to establish the compartments.[17]

Although only a little is known about the circuitry of compartments, evidence supports our assertion that the body plans and corresponding compartment maps are easily installed. In the anterior-posterior dimension of *Drosophila* there are at first only two localized components, one at each pole, that initiate six domains of gene expression in early development. These in turn set up the eight Hox compartments plus several anterior head compartments and compartments at the extreme anterior and posterior termini. They also establish the segmental compartments, each with a distinctive anterior and posterior compartment inside it, repeated 14 times. In the back-to-front dimension are five compartments set up from a single gradient of a transcription factor, which has its high point of activity on the belly side. Thus, the numerous anterior-posterior compartments intersect with five back-to-front compartments to divide the embryo up into roughly a hundred compartmental regions. Thus, rather few localized agents (perhaps only three: one at the anterior, one at the posterior, and one on the belly side) seem to be used in early *Drosophila* development to get the spatially complex compartment plan activated and emplaced. The mammalian egg seems to have even fewer, if any, prelocalized materials.

A second argument for the ease in generating the compartment map comes from mathematical modeling of the regulatory circuitry of the anterior and posterior compartments of the *Drosophila* segment. Are the compartments really self-organizing, or does the fly in its early development have to provide all the components of the circuits at just the right concentrations and locations to get the system started? We have identified each compartment by its expression of a single selector gene, but we have not been concerned with what specifies that expression. In fact, each compartment depends for its persistence on a complex circuitry of secreted factors and transcription factors that cross-activate and cross-inhibit other compartments.

Twenty-eight gene products are involved in the pair of segmental compartments in this circuit. Computer simulation experiments ("in silico") have shown that the 28 components must be connected in one particular circuit of positive and negative interactions to start and maintain the compartments. Other circuits do not work. Once the connectivity is set up, however, the compartments are tolerant of changes in the levels of the 28 components. This scenario means that to get started and to continue, the compartments do not need exact amounts and exact placement of components, and at first do not even need all the components.

The conclusion of the modeling study, when stated as a hypothesis about real development, is that compartments are relatively easy to set up because they are so adept at completing their activation and self-maintenance. We could say that the exact circuitry of 28 components is itself very constrained to change, because of the many specific interactions, but that the constraint has gained for the organism a robustness suitable to easy initiation and maintenance. You do not have to turn the key "just right" to get this motor going; any kick will do. The trade-off for a highly constrained segment maintenance system is, then, a deconstraint on early development.[18]

Putting minimal demands on the egg is probably extremely important in facilitating evolutionary diversification. Although developmental biologists have generally focused on adult morphology, in perspective the adult is merely a vehicle for generating sperm or eggs, providing nourishment to the egg, and protecting the egg. For the female, the production of the unfertilized egg is usually a major investment in energy. In placental animals, the female provides further protection and nutrition, and in some other animals the mother and/or the father broods the egg through early development. Humans provide for their young even through early adulthood. The fitness of an individual is measured by its capacity to reproduce, and the earliest stages of the life cycle, as much as the adult stage, are important for viability.

If evolutionary innovation at the earliest stages of development, well before the phylotypic stage, is critical for fitness, properties of the phenotype that deconstrain innovations in early development could be as valuable as those that deconstrain innovations in late development. A particularly important example of egg innovation is the evolution of the so-called cleidoic egg by early reptiles about 300 million years ago. Before that time, land animals had to return to water to lay their eggs. This restriction limited their exploitation of the earth's expanses of dry land. The changes of egg organization and early development were manifold: a long-term food supply (large amounts of yolk in a large egg) and a means to use it (the yolk sac tissue), a means for gas exchange (oxygen and carbon dioxide) and for handling waste (chorioallantoic membranes), and a means for maintaining an aqueous environment (amnion, shell membranes) under conditions where the ambient temperature might be over 100° F (38° C) and the humidity near zero.

The conserved compartment maps of the different phyla of animals are intimately connected to the diversification of anatomy and physiology of the members of the phyla, and to the diversification of phyla. They provide the means to use different combinations of conserved core processes at different places, in parallel. They create the places at which the different kinds of local development occur. And while complex in their own variety and organization, they have the capacity to be set up easily in development, hence not precluding the egg's simultaneous engagement in other kinds of development related to nutrition and protection of the embryo.

Constraint and Deconstraint

Disclosure of the compartment structure of multicellular animals was at once an insight into embryology and an insight into evolution. It unified our appreciation of how the organism generates the complexity of the adult, and at the same time it gave molecular clues to large-scale and small-scale evolutionary change. The discovery also exposed how unevenly conservation and change are distributed in animal evolution. Compartments are surprisingly more conserved than the anatomical differences that define the phylum, yet the compartment map has to be comprehended as a process operating in the anatomical dimension. It is an example that demands an explanation of the relationship of conservation and change in evolution.

It is only with the concept of the compartment map of the body plan that the notion of higher taxonomic categories, such as the phylum, was shown to be nontrivial. A counterview, expressed by the evolutionary biologist George Williams, was that the body plan might be trivial, a result of "random phylogeny." That is, if everything in the organism changed randomly in evolution, then those animals would be grouped together in which the same 10 percent of characters had by chance not yet changed after some time, say 200 million years. And after 400 million years, those animals with 1 percent of characters still unchanged would be grouped together. There would be no reason for the residual characters to have remained together.

The discovery of the highly integrated circuits that constitute the conserved compartments of the phylotypic stage cannot be explained by random phylogeny. Instead, compartments and the circuits that maintain them represent conserved core processes, rather than bits and pieces of residual related characters. Although the phylotypic compartments might now be so embedded in the development of the phenotype that they are unable to change, we argue that they have also been under continuous selection for the versatility and robustness they provide. Why have they been selected? And for what?[19]

The compartment map does not just record locations in the embryo; it selects the kinds of development to occur at those locations. The phylotypic body plan, with its highly conserved compartments and selector genes, is promiscuous in the kinds of anatomies and physiologies it supports. It is itself not under much direct selection, since it has no differentiated cell types and is still insulated from external challenges by the highly evolved protective conditions of the egg. But it *is* selected along with all the directly selected traits that depend on it for their development each generation.

The robustness of the compartment body plan and its connections by weak linkage to the conserved core developmental processes are key to the facilitation of variety around it. Without weak regulatory linkage, differentiation in the various compartments could not be as diverse as it is. Without compartments, all the flexibility of weak linkage would be thwarted by pleiotropy, the interfering effects of beneficial change in one region causing deleterious change in another. Without compartments, exploratory processes such as axon migration or neural crest differentiation would be limited by the lack of diversity in anatomy. Yet without exploratory processes, diversity would be limited to the cells indigenous to the compartment.

Whenever new anatomical features arise, they must be integrated into the overall anatomy of the animal. The compartments need to talk to one another at the margins, and signals that orient these structures relative to the body plan have to be maintained. The compartment plan continues to operate in this large-scale patterning role without interfering much in local differentiation. Thus, the individual compartment modifications are selected relative to the conserved and unchanging compartment plan.

The compartment plan, because of its anatomical role, is perhaps the most persuasive example of how a robust developmental or physiological process can itself be constrained to change and still deconstrain evolutionary change in other processes. What is the hypothetical alternative to an organism with a compartment map of the kind we have described, but with the same large number of regionally expressed genes in the adult? Roughly estimated from a few animals, a quarter of the genes, or about six thousand in vertebrates, may be expressed in nonuniform patterns in the body, that is, at specific places.

An interesting alternative would be an organism with no compartments. In this organism, six thousand independent pathways would lead from the egg toward the adult's six thousand regional gene expressions, operating in parallel, without interference. Detailed spatial information would be needed from the start for all these lines, making the egg as complex as the adult, with six thousand localized agents. Not only would this network be complicated to arrange, there would be a high vulnerability to conflicting demands on the use of the same gene in multiple places. The evolution of new anatomies and physiologies would require manifold changes of gene expression.

Such an organism by comparison to the compartmental animal would seem less able to evolve complex anatomies with the same input of random mutation. The conserved compartment map seems an efficient solution to the problem of spatially patterned expression of six thousand genes, many of which are expressed in several compartments. That is to say, early development emplaces the compartment map of perhaps one hundred (*Drosophila*) or two hundred (chordates) expressed selector genes; the encoded products of these genes emplace the remainder of the six thousand target genes. Furthermore, the robustness of the process of development is remarkable in light of the variability in cell number and cell placement, and in the face of environmental stress. Further study is needed, perhaps on a theoretical level, to understand how compartment organization actually provides this robustness and whether it offers demonstrable advantages to the alternative strategy of detailed prelocalized information.

Extending Compartment Thinking

Compartmentation as a strategy is one way of generating complexity from a relatively small number of genes and of avoiding conflicts due to use of the same gene product in more than one context. Although we have only considered compartmentalization in the context of spatial organization in the embryo, other complex processes are compartmentalized in different dimensions.

One other dimension is use of the single genome of an animal in many stable expression combinations, say the three hundred differentiated cell types of a complex animal such as a vertebrate. Each cell type, because of the different genes it expresses, contains a different profile of proteins and RNAs, and thereby a different appearance and function—which is its cellular phenotype. The current hypothesis, with solid evidence in several cases, is that each cell type is distinguished by the expression of one or more master regulatory proteins (transcription factors) encoded by a so-called master regulatory gene that is continuously expressed due to a *positive feedback loop*. That is, the gene is directly activated by the protein it encodes, once its expression has started. The master regulator, like the selector protein of a body plan compartment, then activates or represses many target genes, determining the profile of RNAs and proteins unique to that cell type.

Unlike selector proteins, the master regulatory proteins are not linked to the overt spatial dimension of the body, but they do bring together a combination of expressed genes in one cell, and hence a combination of conserved core processes. They select one combination of expressed target genes from all the possible combinations of expression of the 22,500 genes of the genome. A cell type is a compartment in "gene expression space." As in spatial patterning with Hox genes, the master regulators can put a subset of the genome under specific control and limit the behavior of that set of genes so that it requires the action of the master gene.

We can extend compartment thinking to dimensions of time and populations as well. With respect to time, the life cycle of an animal can be divided into a series of stages, each of which can be specialized with different anatomies, physiologies, and behaviors (as in the larva, juvenile, and adult). Often these stages are specialized for life in different environments, so the animal is a serial specialist. The alternative would be to have one complex animal form, a generalist, with all the anatomies, physiologies, and behaviors simultaneously manifest.

The argument is that the compartmentation in time allows greater specialization of each stage, because conflicting demands are reduced. In this manner, a larva specialized for one niche can metamorphose into an adult specialized for a different niche (Chapter 3). We do not know whether each stage of the life cycle is compartmentalized by a master regulator protein, either a single one or a circuit of them. In the nematode *Caenorhabditis elegans*, this may indeed be the case: proteins and RNAs encoded by so-called heterochronic genes activate and repress other genes of each of the four larval stages. Loss of each such gene means loss of one larval stage, with repetition of the previous stage, a homeotic transformation in the time dimension.[20]

The innovation of compartments in the dimension of space, or cell type, or time, or sex, or phenotype is that each is distinguished by at least one unique transcription factor or unique combination of factors. All other gene expressions are entrained by this factor or factors. In the case of the compartment map, each compartment produces signaling proteins as well. Gene expression in a compartment would have to function as a logical AND system. (Logical AND systems require all inputs to be active to obtain an output, whereas logical OR systems require only one of the inputs to be active.) In compartments, some aspect of any given gene's expression would be dependent on

the continued presence of the compartment identifier, its selector protein "tag."

This system is simplest to see for the case of a gene product that is unique to a compartment. A larval gene might be activated by a selector factor specific to the larva, or a gene for egg production might be activated by a female-specific factor. In this view, the target gene would require such a specifier (the selector protein or master regulator) for its expression. Alternatively, a certain gene could be repressed in a specific compartment, in which case the specifier would inhibit expression. More complicated and perhaps more commonplace, the specifier might modify the gene, or modulate its level of output, or generate a specific splice form, or force the gene to require yet another environmental signal.

The invisible anatomy of the embryo is much more than a new set of anatomical features that require special methods for visualization. It is a compartmentation of the organism that allows independent evolution and development, while maintaining overall coordination of activities related to the body plan of the phylum. It enables the relatively small genome to be used in hundreds of different ways and avoid interference of one region with another caused by conflicting needs for the regulation of gene expression. The spatial compartments represent only one form of compartmentalization, but each offers the same benefits: an allowance for simultaneous highly adapted uses of the biological system—or more specifically the genes, the developmental programs, or the different parts of the anatomy—in place of more generalized and more complex uses.

These compartmentalized systems are themselves very complex circuits. The highly constrained compartments that make up the body plan might be expected to require fragile developmental processes, where everything has to be organized precisely in the egg for the compartments to be set up and for subsequent development to take place. It might have resulted in a system so complex and interconnected that it would be difficult to make viable changes in evolution at the egg stage. On the contrary, the developmental circuits, supported by weak linkage and the exploratory mechanisms that make up the body plan, constitute a system that is tolerant of change in the processes that precede and emplace the circuit of selector genes at the phylotypic stage and is permissive of modification that follows the phylotypic stage. Compartmentalization involves some of the most constrained and conserved circuitry in biology, but as a trade-off it achieves deconstraint and robustness in evolution and embryonic development.

SEVEN

Facilitated Variation

We summarize here in explicit terms our theory of facilitated variation, distinguishing it from other theories and providing evidence for it. The theory rests on molecular knowledge of a host of conserved cellular and developmental processes, which underlie and connect both somatic adaptation and phenotypic variation. Many evolutionary biologists do not see a need to connect somatic adaptability to the generation of variation, and some see a need to keep them separate. For them, it is sufficient to say that random mutation is required and that phenotypic variation arises haphazardly from it as random damage; the organism's current phenotype does not matter for the variation produced, and the output of variation is nearly random. But from the research advances of the past few decades, it is apparent that biologists have underestimated the range of somatic adaptations produced by conserved processes, especially those operating within the animal during embryonic development. Our theory of facilitated variation puts the organism and its pervasively adaptive phenotype at the core of the process leading from random mutation to nonrandom phenotypic variation.[1]

Previously we, and others, have considered the question of the organism's capacity to expedite its own evolution under the term *evolvability*. If we define evolvability as the capacity to evolve, it is but a tautology. It gains meaning when resolved into a variation component and a selection component. If an organism had been fortuitously preadapted, from previous selections, to deal with a new environmental threat or opportunity, its evolvability might be very great but it would imply no specific features of its biology or construction. A lineage that evolved quickly and filled many niches would retrospectively have been highly envolvable; but that may only reflect the coincidence between special features of the environment (the selection component) and special properties of the organism.[2]

It is the variation component that has been the focus of this book. Here we can talk about the organism's capacity to generate phenotypic variation in response to genotypic variation and about the nature of that variation, independent of variations in the environmental conditions. We can talk about the design of the organism, its core processes, and their special characteristics in suppressing lethality and in providing a quantity and quality of phenotypic change. These features can, in principle, be measured in an individual organism each generation, though their consequences for evolution can only be assessed in a population over time. For that reason we have formulated our theory of facilitated variation to deal with the variation component of evolvability alone; we relegate to fuller treatments of selective conditions an evaluation of the capacity of a population to evolve.[3]

The Variation Component of Evolvability

We explain the variation component of evolvability by our theory of facilitated variation. Aspects of the theory have been developed throughout the book, both its historical roots and its grounding in modern cellular and developmental biology. We now outline the intact theory of facilitated variation:

1. Despite the randomness of mutation (with respect to selective conditions), phenotypic variation cannot be random because it involves modification of what already exists.
2. The existing organism constrains and deconstrains variation of its phenotype, both the kind and amount. Some

components and processes are constrained in the change they can undergo, but deconstrain the change of other components and processes of the phenotype. The overall trade-off is such that phenotypic variation is accelerated over what would occur if deconstraint were absent.

3. Variation from this trade-off is both less lethal and more appropriate to selective conditions than would be variation from random damage. Evolutionary change is thereby facilitated.
4. The constrained parts of the organism are the *conserved core processes.* Each process involves several protein components, conserved in their sequence. Their function is to generate the phenotype from the genotype. These processes arose historically in a few intermittent waves of innovation. On the lineage toward humans, these innovations include the processes of the first bacteria, of the first eukaryotes, of the first multicellular organisms, of large bilateral body plans in metazoans (including chordates and vertebrates), of neural crest cells in vertebrates, of limbs in the first land vertebrates, and of the neocortex.
5. The core processes have been remarkably unchanging over time. For example, the basic information processing of DNA, RNA, and protein synthesis is the same in all living organisms; the functions of intracellular membranes and the cytoskeleton are the same in eukaryotes; the functions of junctions and the extracellular matrix are the same in all metazoa; the developmental role of the Hox genes is the same in all bilateral metazoan phyla; and the developmental program for limb formation is the same in all land vertebrates.
6. Most evolutionary change in the metazoa since the Cambrian has come not from changes of the core processes themselves or from new processes, but from regulatory changes affecting the deployment of the core processes. These regulatory changes alter the time, place, circum-

stance, and amount of gene expression, RNA availability, or protein synthesis of components of the core processes, or alter the activity and interaction of proteins of the processes by modifying them or by changing their stability. Because of these regulatory changes, the core processes are used in new combinations and amounts at new times and places. Also because of the regulatory changes, different parts of the adaptive ranges of performance of the processes are used in new circumstances.

7. Protein evolution, which has occurred extensively at those rare intervals when new processes emerged, is itself an example of conserved core processes at work. It involves genetic recombination, the exon-intron structure of genes, and RNA splicing to generate new composite proteins from existing coding domains.
8. Physiological processes that adapt the individual to environmental conditions are rich targets for evolutionary modification. These processes contain combinations of conserved core processes. As J. M. Baldwin, I. I. Schmalhausen, and others have suggested, evolutionary change can in some cases simply entail the displacement of an existing physiological range by external conditions, followed by mutation to stabilize and enhance the marginally adapted state.
9. We propose that a much richer source of targets is to be found in the conserved core processes of development and cell behavior, the processes directed inside the organism rather than toward the environment. Each of these processes has a physiological range and an adaptive potential, selected for the robustness conferred on embryonic development and adult physiology. Stabilizing these core processes outside their normal range can generate new phenotypes that are already integrated into existing developmental events. Such regulatory modification of existing processes is likely to be less lethal and generate more

phenotypic change, for an input of random mutation, than would be gained by inventing new structures or physiology.

10. The core processes are built in special ways to allow them to be easily linked together in new combinations, and to be used at new times and places, to generate new phenotypes. These special properties include:
    a. *Weak linkage, a property particularly of signal transduction and transcription.* In weak linkage, the protein interactions are weak and indirect. The signal is minimally informative and not instructional, whereas the response is maximally prepared and ready to be triggered. Weak linkage usually implies that a preconditioned response, which is self-inhibited, is released by the signal. With this prior preparation of responding components, demands on the signal for precision of interaction are low. Weak linkage facilitates the evolutionary change of signals and combinations.
    b. *Exploratory behavior, a property of the processes forming the cytoskeleton, of processes operating in cell groups in development, and of functioning populations of organisms.* The exploratory process has the capacity to generate an unlimited number of outcome states. Then, in response to an input, one or a few outputs are selected from among those states and retained, often by stabilization. Since only one state is eventually used, the unselected states are nonfunctional under the specific conditions. Yet these nonfunctional states may gain roles in future evolution. The selective agent, in its regulatory role, does not have to inform the process of its outcome. The process is built to be receptive to the agent, which simply serves as a stabilizing force, selecting one state among the large number of states generated

in each instance. Therefore, new selectors are easily originated and with them new outcomes.

c. *Compartmentation, a property of embryonic spatial organization and cell type control.* Compartmentation involves the use of weak linkage in the spatial dimension to specialize the behavior of different genes and different processes in different topological domains in the embryo. The domains allow a largely independent evolution of different regions of the animal, a property that has facilitated a large increase in the complexity of anatomy and physiology of animals without a corresponding increase in the complexity of the conserved core processes.

11. The generation of variation is facilitated principally by: (a) reducing pleiotropy, the lethality of mutations in one part of the phenotype that might have selectable benefits in another part, (b) increasing the amount of phenotypic change gained for a given amount of mutational change (or, said in the reverse, reducing the number of mutations needed to produce novelty), and (c) increasing the genetic diversity in the population by suppressing lethality.
12. Our theory of facilitated variation, stated here in its most coherent and complete from, gives what we think is a plausible account of the dependence of phenotypic variation on genotypic variation, indicating that novelty mostly draws on what is already present in the phenotype, and further indicating the role of conserved components and processes in innovation. The theory is new in its attempt to map cellular and molecular changes to evolutionary events by drawing on the molecular parallels among physiology, development, and evolution. The facts used to support our arguments, though mostly recent, are widely accepted. It is their application to the problem of phenotypic variation that is new, and we believe it to be of great explanatory power in completing Darwin's theory.

Robustness, Flexibility, and Versatility

Robustness is a general property of the phenotype, related to the adaptability recognized by previous authors. We have placed it at the center of our theory of facilitated variation and tried to explain it on a cellular and molecular level. Robustness can only be evaluated comprehensively by looking at the whole organism, since the fluctuations of individual processes in the organism may compensate for one another. The properties of compartmentation, exploratory behavior, and weak linkage reduce the interference that might be expected if modified genes were to affect many processes at once. Furthermore, the conserved processes are built with various feedbacks, self-adjustments, and compensations to give sufficient outputs despite altered conditions and inputs.

All the forms of robustness, flexibility, and versatility are expected to enable conserved processes to work together, and to be brought together, in different combinations, under the various conditions met at different times and places within the individual organism. These characteristics also lead to buffering against genetic damage, and hence a storage of genotypic variation in the population of animals, as was recognized in the heating experiments of Waddington and Lindquist. At the same time, robustness affords tolerance to evolutionary changes involving use of a process in new combinations, times, and places, all of which would upset a nonrobust process and render it lethal.

It may seem counterintuitive that robustness should make it possible for small amounts of regulatory change to unleash a large variety of evolutionary changes, especially in the realm of anatomy and physiology. George Gaylord Simpson dismissed somatic plasticity and adaptability in the Baldwin effect as unimportant in evolution because the Baldwin effect only drew on the not-very-different physiological and anatomical states already present in the organism and therefore would be irrelevant to the generation of real anatomical novelty. All the same, extending some of Baldwin's insights to the internal adaptabilities of core processes, as understood in the 1990s, counters Simpson's skepticism. The example of the neural crest addresses phenomena close to Simpson's morphological tradition. We showed how exploratory cells that migrate over the complex variety of compartments of the body plan can generate many new anatomical features based on their broad responsiveness to local signals.

We suggest that when these conserved processes accumulated in evolution and gained their properties of robustness and flexibility, the organism became more and more a system capable of responding to random mutation and other forms of genetic variation. It did so by using existing processes to produce phenotypic variation, via regulatory changes. Organisms as they evolved did not improvise each phenotypic change independent of all that had gone before.

What we are describing is truly descent with modification. In the hypothetical evolvable organism, richly endowed with a capacity for facilitating variation, a small input of random mutation would lead to a large output of viable phenotypic variation. In cases where the effect of mutations is well buffered by the adaptability of these same processes, there might be no change in phenotype. Even where buffering dampens phenotypic variation, evolvability might also be served. The population would simply accumulate these genetic changes, and that reservoir of changes could serve to rapidly generate variation under some future conditions.

Instead of a brittle system, where every genetic change is either lethal or produces a rare improvement in fitness, we have a system where many genetic changes are tolerated with small phenotypic consequences, and where others may have selective advantages, but are also tolerated because physiological adaptability suppresses lethality. The impressive tolerance of substantial phenotypic differences is seen in the viable dog mutts that arise from very different parental stocks.[4]

In the generation of phenotypic variation from random mutation, the organism as a whole is not a blank slate but a poised response system, rather like one of the signal-response systems within its physiology. It responds to mutation by making changes it is largely prepared in advance to make. Its adaptive envelope of responses is far greater than that which can be elicited by testing every environmental condition on the whole organism, as Schmalhausen would have us do. The envelope includes all responses possible within the cellular and developmental processes of the organism, and some of the exploratory processes have an infinite number of responses. Simpson was right to think that the organism cannot express every morphological variation as a somatic adaptability before it can co-opt the variation for a heritable anatomical innovation. These variations may in fact require new mutation or reassortment of existing genetic variation in the population. Genetic variation or mutation does not have to be creative; it only needs to trigger the creativity built into the conserved mechanisms.

Facilitated variation, then, has two sides. It is a combinatorial theory, where the elements that are combined are individual functional processes. Weak linkage and exploratory mechanisms allow various processes to be linked with new inputs connected to new outputs. It is also a state-selection theory, where regulatory changes evoke parts of the adaptive ranges of the processes.

We have seen such a selection of parts of the reaction norm in physiological examples such as hemoglobin or in developmental examples such as the omission of the larval stage in direct-developing sea urchins. It has been convincingly argued that entire developmental circuits have been moved around in parasites, salamanders, and insects. From new combinations we derive whole new physiologies and anatomies, and from state selection we also obtain new physiologies and anatomies, ones that have already been tested. Novelty comes from these two sides and from their interaction.

Regulation of Gene Expression

A modern view sharing some features of facilitated variation is the *cis*-regulatory model of evolutionary change. Some scientists now working with the genome and with gene expression in developing animals take it as a plausible and sufficient model of evolution. The term cis-*regulatory* refers to the DNA sequences that are adjacent to a gene, through which the gene's transcription is controlled (*cis* is a Latin preposition meaning on this side of).

According to this attractive view, the most important evolutionary change is that occurring in the regulatory regions of genes, by which random mutation creates or eliminates sites in the DNA for binding various transcription factors of the great variety available in the cell. Each site is a few bases in length, perhaps six to nine, and not entirely unique in sequence since some positions can carry alternative bases. When a site changes and is newly bound by a factor, the expression of the gene changes its time, place, or amount, depending when and where that factor is present. This change of regulatory sequence is, of course, heritable. Thus, without too many special requirements, a new condition for expression can be added without losing the old. Many candidate factors are active in binding to DNA only when the cell receives external signals or when the factors are carried forward in the cell lineage from earlier developmental stages. The newfound transcription of the gene therefore relates to some spatial-temporal aspect of development of the embryo.[5]

The changing of *cis*-regulatory DNA by mutation does not affect protein structure in most cases, because *cis*-regulatory sequences are not usually transcribed or translated into protein. Since changes of sequence in the *cis*-regulatory region do not cause detrimental amino acid changes, they are not eliminated by selection as a result of functional failures of proteins. Beyond this, improved binding sites would be preserved by positive selection. This model provides a direct and efficient means to express genes in new combinations and amounts, with little investment of mutational change—exactly what our facilitated variation theory seeks. Transcription with its means of regulation embodies these conserved features of ready modification and is one of the most important of the core processes that manifest weak linkage.

Yet this model falls short as a complete description of phenotypic variation. We understand this *cis*-regulatory model as functioning consistently within the theoretical structure of facilitated variation that we have described, but as only one part, a subset, of our larger theory. The *cis*-regulatory model implies little about what happens to generate actual phenotypic variation. It would have to say more about which genes are regulated, that is, about the transcripts and proteins. The model is consistent with the predominance of conserved proteins in organisms, but is silent on the role of these proteins. Conservation is implicit because the change of coding sequences in DNA is not considered. The model implies that the proteins make development happen, however that is done. It assumes that the linkage of gene expression in new combinations through transcriptional regulation will be effective in generating new kinds of development and hence new traits.

The proponents of *cis*-regulatory models have given little attention thus far to the conserved components and processes, and to the special properties that ensure their working together in different combinations and amounts. We, on the other hand, have expounded the kinds of processes and special properties that enable their versatile usage; we have focused rather little on the details of regulation. We feel, in fact, that all the special properties of the conserved processes had to evolve before regulatory evolution could escalate, for if the components of different processes were to interfere with one another in the new combinations, such expression would afford no benefit. Thus, while we too give a central place to *cis*-regulatory evolution, we surround the center with the context of what is accomplished in phenotypic change. We also place in the center all changes of coding sequences of genes for regulatory agents and for regulatable agents that can change by random mutation and reassorted genetic variation.

The cell has many ways other than transcriptional *cis*-regulation to control the presence or absence of protein function, and thereby to control the time, place, or amount of function. Transcription factor proteins are themselves targets of change. Many contain repetitive runs of a few kinds of amino acids. These runs expand and contract at high frequencies, many thousands of times higher than other mutations, with the consequence of increasing or decreasing the factor's activity.

Some of these changes correlate with skull shape in various breeds of dog; another change, entailing a loss of 17 repeated amino acids in a transcription factor present in limbs, correlates with the presence of a second dewclaw on the rear leg of the Great Pyrenees breed, as shown in Figure 35. Furthermore, the addition of such a repeat sequence to one of the Hox proteins, as well as the loss of a sequence for receiving phosphate signals, correlates with a newfound suppres-

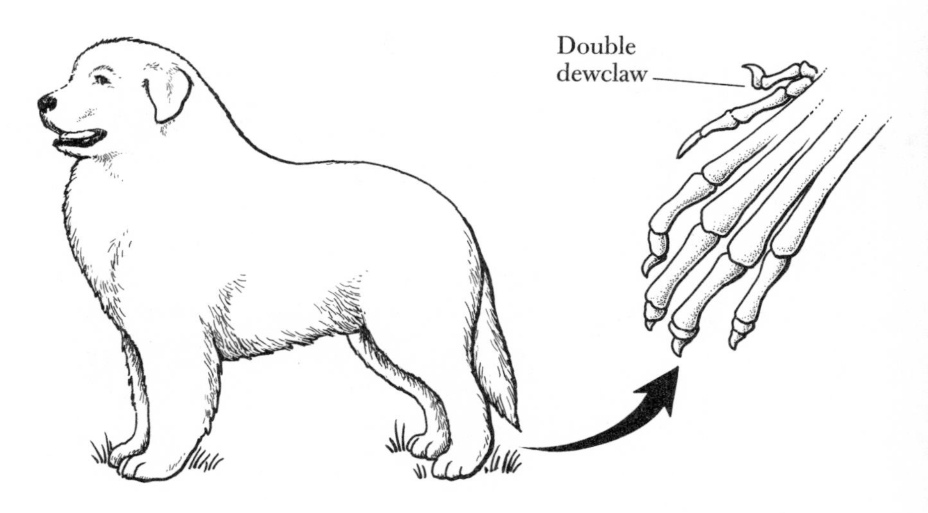

Alx of other dogs

-FPPQPQPQPPAPQQPQPQQPQPQPQPPAQPPHLYLQRGA-

-FPPQPQPQP-----------------------------PAQPPHLYLQRGA-

Alx of Great Pyrenees

Figure 35 The rapid evolution of dogs. The Great Pyrenees breed has a double dewclaw on each rear leg—an extra pair of digits—which is a unique trait of the breed. The Alx-4 protein of Great Pyrenees is 17 amino acids shorter than that of other dog breeds. In all dogs this protein is present in the hind legs.

sion of appendages in the abdomen of insects, whereas this change did not occur in the many-legged crustacean-like ancestors. Both of these are examples of changes in coding sequence for new forms of functional regulation, not changes of *cis*-regulatory control. In contrast to the *cis*-regulation hypothesis, facilitated variation is intended to explain comprehensively how a minimal input of random mutation can generate phenotypic variation.[6]

Evolutionary Change and Facilitated Variation

Now that we have laid out our theory, it will be useful to return to the dominant theories of evolution to show what facilitated variation adds;

how is it new? Let us examine scenarios comparing morphological innovation in Baldwinesque terms, in neo-Baldwinesque terms as developed by Schmalhausen, Lindquist, and West-Eberhard, and in neo-Darwinian terms. We will consider these scenarios first without and then with the contribution of facilitated variation.

A reasonable example would be the rather rapid evolution of bird beaks among the Galápagos finches. Darwin was struck with the great variety of beaks, and they were investigated over several decades in the late twentieth century by Rosemary and Peter Grant, both ecologists and evolutionary biologists, whose work was memorialized in the Pulitzer Prize–winning book, *The Beak of the Finch*. The general history of the Galápagos finches is well understood. In the space of a million years or fewer, a founding group of finches from the South American mainland generated several species on the islands, some with large, pliers-like beaks for cracking large nuts and some with forceps-like beaks for extracting insects from fruit, as illustrated in Figure 36. On the evolutionary timescale, beak shape appears to be plastic. Neural crest cells are central to beak development, so we may expect to find their adaptive cell behavior again in this example.[7]

Through Baldwin's Eyes

As our evolutionary scenario, we imagine Baldwin arguing from the general position of somatic adaptability. He might explain the rapid radiation of the founding finches into different species by saying that as the climate changed and the food supply shifted from nuts toward fruit, the finches with large beaks would be increasingly stressed to their adaptive limit. As a somatic adaptation, they would develop smaller, thinner beaks. Little is known about the adaptability of beak development, but we could imagine that juveniles, trying to stick their beaks into small holes, exerting less force on the softer food, or growing up on the edge of starvation, might adaptively develop smaller beaks. After all, a human bone can develop thickly or thinly depending on load and nutrition. Some birds, even a significant fraction of the population, might just manage to extract insects from fruit adequately but not well, for they are at the limit of their adaptive capacity. Nonetheless,

Figure 36 Darwin's finches reconsidered. The beaks of two species are shown, as first observed in the Galápagos Islands in 1834. The thick-beaked species specializes in eating nuts and hard food; the thin-beaked species, in insects, fruit, and soft food.

they survive and reproduce marginally. Opportunities for selection of birds with improved reproductive fitness would always exist. Stabilizing mutations or genetic reassortment would arise in some birds in the population. The result would be beak refinements, in the course of which some birds would heritably produce forceps-like beaks even when the climate and food variety changed to the renewed prevalence of nuts. The adaptive range of beak development would be permanently shifted by the stabilizing mutations toward small size, and the finches might become incapable of cracking large nuts at all.

The Baldwin scenario is most plausible when there is a known, existing physiologically adaptable process, directed toward external conditions that can be stabilized. For temperature, salinity, diet, oxygen tension, or resistance to environmental toxins, we can assume these physiological adaptations exist. However, we have no reason to think that there is a process for beak development, influenced by use, that would produce a large range of anatomies—from beaks with the ability to extract small insects to beaks with the ability to crack large nuts. The general lack of physiological responses leading to anatomical innovation is the reason that most evolutionary biologists agree with Simpson that the Baldwin effect would work in only a few special cases where evolutionary change and physiological variation were coincident.

Through the Eyes of Schmalhausen

In the neo-Baldwin scenario, we would propose that beaks changed by way of morphosis, as suggested by Schmalhausen and explicated by West-Eberhard; these views were supported by experiments of Waddington and of Lindquist. Environmental conditions might change, for example, to a hot dry climate. Some of the heat-stressed birds might exhibit aberrant development, producing thin beaks that would have no direct adaptive value in offsetting the climatic stress. The thin beak might be the kind of phenotypic change evoked by Susan Lindquist with heat shock. However, thin beaks would be fortuitously valuable for the fruit-insect diet, which coincidentally is more available under these conditions. As in Baldwin's scheme, a significant fraction of the population might produce such beaks, and the persistent hot dry climate would ensure the recurrence of thin beaks over a run of generations. During that time interbreeding of finches would bring out a refinement and stabilization of such beak development, reduced stress, and improved reproductive fitness, eventually leading to a heritable development of thin beaks with less or no stress dependence.[8]

The plausibility of this scenario depends on the likelihood that exceeding the physiological range and generating aberrant morphologies would produce a phenotype that would be functional and serve an unrelated purpose. While it is true that the morphological variations in the Lindquist and Waddington experiments were not simply monsters or disorganized tumorous tissue, it is not true that they were necessarily adaptive. Waddington opportunistically chose particular morphological variants that correlated with heat and ether shock, namely, wings lacking cross-veins and four-winged flies. There is no assurance that he would have seen, at appreciable frequencies, any particular adaptive morphology. Furthermore, for the case of the finch beak, these morphologies would not only have to be produced at a significant frequency but would have to be integrated into the developmental program of head formation of the bird. Absent any understanding of the mechanism of beak morphogenesis, it is unlikely that any given adaptive outcome would appear due to an unrelated stress.

The View from the Modern Synthesis

A standard neo-Darwinian explanation might have the beak phenotype of the founder birds rather fixed and narrow, with no developmental adaptability of the beak to food hardness or softness. Mutation, recombination, and reassortment of genes in the population would occasionally yield variant birds that had heritably, smaller and slightly more forceps-like beaks. These would have some success in exploring a different food niche (insects in fruit), which might also be geographically separate from the hard-nut food sources of the main population. Each mutation would bring a new element to the beak phenotype; change would be gradual, but repeated selection of better-adapted rare variants would drive the population in the direction of a forceps-like beak and a well-adapted, insect-extracting species. Since beak size would need to vary independently of most other traits, it might be assumed that the mutated genes would be exclusively involved with beak morphogenesis.

This explanation founders again on our lack of understanding of beak development. How many separate parameters does it take to specify a beak, and would each of these parameters be regulated by a separate gene or multiple genes? To develop the size and shape of the beak, would the modifications of each gene product have to be small so that at each stage the bird would be left with a functional morphology? For example, the upper and lower beaks, which derive from different founder cells, would have to be coordinated in their growth. In many ways the Darwinian view of evolution is like a movie, which looks continuous but is made of many frames, each of which is an abrupt, albeit small, shift from the previous frame. In this case, the shifts are not only small but also uncoordinated, and in fact they have to be small so that a rough coordination appears at every step.

If this is the model of beak morphogenesis, then a sequence of many very small changes is necessary, each supported by a mutation, each selected in the right sequence, and the entire process occurring rapidly. Issues of the pace of evolutionary change enter here, given the low frequency of mutation, the relatively long generation time, the small populations, and the requirement that the very small changes be selectable.

The View from Facilitated Variation

Facilitated variation would enable all three points of view. The plausibility of each would be different depending on the anatomical process, but under some conditions we would expect any of the three scenarios to be feasible when incorporating features of facilitated variation. The size and shape of the beak are, we know, determined by the growth and differentiation of five small nests of neural crest cells that settle around the mouth of the embryo. The nests receive signals from facial cells at the five sites and respond to them. Thus any features that affect neural crest cells would affect beak growth coordinately.

The change in shape of the beak reflects the differential responses of these five populations. The extent of proliferation in each nest is one kind of response, and the quantity of hard-beak material each cell deposits is probably another response. We can imagine scores of factors that could control such responses, and that have a quantitative scale to them. They would include the initial number of neural crest cells in a nest, competition from other sites, the level of secreted signals that stimulate or inhibit proliferation, and the level of secreted signals that induce cell death, or that induce differentiation earlier or later, more or less.

Each of these factors could be controlled at the level of transcription, splicing, RNA stability, protein and RNA transport, protein modification, and protein degradation, of any of a large number of genes and gene products controlling proliferation and differentiation. The decision of the neural crest cells to divide is simply a balance of many inputs, as is the differentiation of beak material. Regulatory opportunities abound for quantitatively modifying the process. The basic program of beak development would not have to change at all.

This view for explaining beak change would emphasize that a few regulatory changes, based on heritable mutational changes, could select on the wide range of outputs available to these adaptive and versatile cells. These changes could stabilize a limited part of their extensive adaptive range to generate the novelty of beak size and shape. The adaptive cell behavior of the neural crest cells of the beak may or may not be used by the bird for a somatic adaptability to hard and soft food. Nonetheless, the adaptive cell behavior of neural crest cells and the facial environment is always available as part of the mode of development. In light of the minimal novelty involved and the wide adaptive range of the neural crest cells, it would seem possible that rather large changes in beak size and shape could be accomplished with a few regulatory mutations, rather than a summation of a long series of small changes.

Finally, the process of beak modification has been well explored in birds, and we can imagine that existing pathways assure coordinated development of the beak and the head. These pathways would provide physiological robustness to variation in any of several gene activities. Therefore, modification of these pathways by adjusting the level or the duration of any component would most likely be integrated into a functional developmental program. The likelihood would escalate that environmental stress or random mutation would be interpreted by the system to give a functional outcome, increasing the feasibility of the neo-Darwinian and neo-Baldwinian scenarios.

We emphasize that our theory of facilitated variation does not replace these earlier theories, but rather complements and completes them. All three theories are possible; the first two, including the morphosis alternative, concern the path of change more than its fundamental feasibility. Along with Mary Jane West-Eberhard, we emphasize the role of adaptable processes within the organism, as opposed to the organism's environmental adaptability. Questions still to be answered are, Which scenario is most plausible in terms of what we know about cellular and developmental processes? What is the likelihood of particular kinds of changes imputed to generate a rather rapid change in beak shape? Which view provides an explanation and a context for many of the discoveries of modern biology?

The Evidence

The most promising experimental path for revealing facilitated variation may lie in studies of the evolution of development, that is, comparisons of the developmental processes of different animal groups and analysis of genetic changes associated with their differences. The aim of such research is to work out what has actually changed in the development of traits of different groups of organisms (beaks, limbs, fins, bristle patterns, color patterns), to identify the conserved processes involved in their development and function, and to identify the regulatory modifications bringing them together in the trait and setting the output ranges. Eventually, we will learn what heritable regulatory changes have been selected in the line of ancestors. This knowledge will yield estimates of how hard or easy it has been to generate a new trait in terms of the numbers and kinds of mutationally effected changes.

Although a large number of differences of DNA sequence are known to exist in the genomes of even closely related species, it is expected that a much smaller number of differences is important in generating phenotypic differences. But which ones? At some point, such heritable regulatory changes will be created in a test animal in the laboratory, generating a trait intentionally drawing on various conserved processes. At that point, doubters would have to admit that if humans can generate phenotypic variation in the laboratory in a manner consistent with known evolutionary changes, perhaps it is plausible that facilitated variation has generated change in nature.

Such experiments are just now becoming feasible. Clifford Tabin,  a developmental biologist who has made major contributions to our understanding of vertebrate limb embryology and evolution, has along with members of his laboratory examined beak development in 6 of the 13 species of Darwin's finches from the Galápagos Islands. They find that the embryos of the various species differ in beak development in a way correlated with the level in the beak of a certain growth factor protein, called Bmp4, which stimulates the deposition of bone (and probably beak materials). Neural crest cells produce Bmp4 in the beak region. In the large Galápagos ground finch, Bmp4 is produced earlier and at higher levels than in the pointed small-beak species.

If this factor is experimentally introduced into the beak neural crest cells in chicken embryos, they develop broader, larger beaks than normal, similar to the beaks of the ground finches. Other growth factors do not have this effect. Nonetheless, when the experimental beak changes size and shape, it is still integrated into the anatomy of the bird's head. It is not a monstrous aberration. The precise regulatory changes accompanying the changes of Bmp4 production in the finches still must be established.[9]

Significantly, what changed was the time and level of expression of a highly conserved signaling molecule, BMP4. It is found in all metazoans, even jellyfish. Changing the level artificially in a different species, the chicken, leads to similar effects. What seems to happen is exactly what might have been predicted by facilitated variation. The changes are regulatory, affecting time, place, and amount of Bmp4. They perturb in a quantitative way via Bmp4 the adaptive cell behavior of neural crest cells, themselves a conserved multipurpose adaptive agency of development. Crest cells do not produce an outright deformity, but in fact modify their development compatibly with the rest of the head.

It would be interesting to examine other examples of break development (toucans, hornbills, hummingbirds) to see if their differences reflect perturbations of conserved core processes and to determine whether similar outcomes in different birds can be generated in different ways. Such information would shed light on the ease or difficulty of achieving phenotypic variation in such systems.

Another example is the anatomical differences of various subspecies of the three-spine stickleback fish. Several subspecies have undergone a relative reduction and bending of the large pelvis, as illustrated in Figure 37, and geological evidence suggests that this alteration occurred in fewer than ten thousand generations (approximately ten thousand years). Pelvic reduction may confer advantage to bottom-dwelling species.

The mutational changes have been mapped by matings of inter-

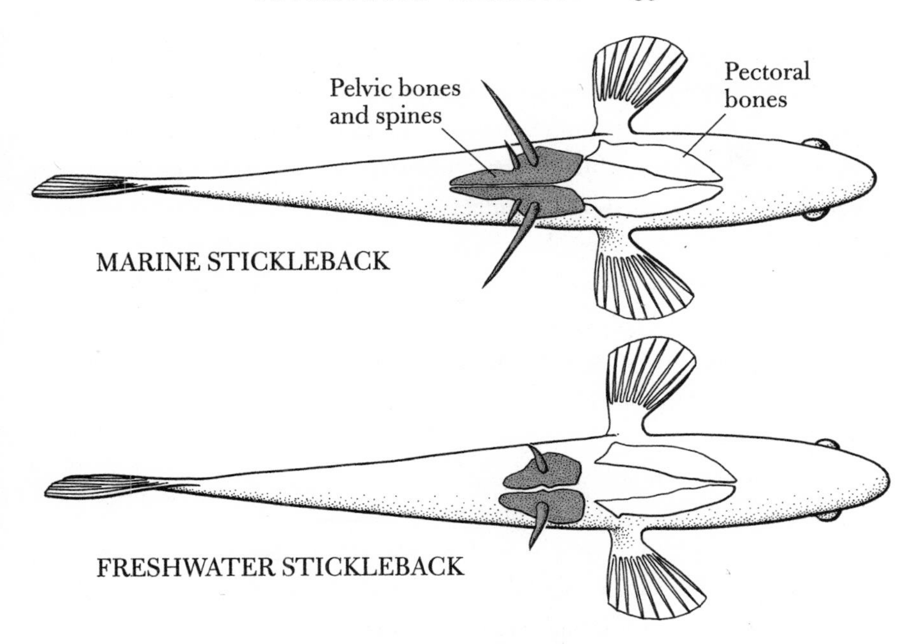

Figure 37 Mechanisms of evolution in stickleback subspecies, which live in salt water or lakes of western Canada. The large-pelvis species (*upper*) occupies open surface water; the small-pelvis species (*lower*) is bottom dwelling near shores.

breeding sister subspecies. This mapping has led to the identification of a regulatory change in the expression of the gene for a transcription factor Pitx1, which is highly conserved among all bilateral animals. Pitx1 protein is involved in many developmental events, including the generation of left-right asymmetry of the whole animal, cranial facial anatomy, pituitary development, and heart development. In the case of pelvic reduction in the particular stickleback subspecies, Pitx1 expression is lost only in the pelvic region. It is normal everywhere else. The specific loss in the pelvis is due to a change in the *cis*-regulatory DNA sequence close to the gene. It is no longer activated by a selector gene product unique to a compartment of the pelvic region. Compartmentation is a critical feature, then, in calling forth the Pitx1 protein for normal anatomical development of the pelvis.[10]

Weak regulatory linkage by transcriptional regulation, depending on *cis*-regulatory DNA sequences, is an easy way to call forth Pitx1 locally in normal development, and an easy way to lose it in the subspecies. Pelvic reduction has occurred on multiple occasions in fish and various terrestrial animals. It will be interesting to see how many different regulatory changes have led to the same loss of localized Pitx1 expression. What is important is that developmental processes accommodated to the local loss of Pitx1, and viable fish of highly modified anatomy were generated. We have previously cited evolutionary changes associated with the change of protein coding sequence; witness the transcription factors in dog breeds (Figure 35) and Hox alterations in insects.[11]

A different kind of evidence of facilitated variation can be found in evolutionary convergence—the evolution of similar organs in different animals that could not have had a common ancestor with the organ. Reptiles and fish have evolved placental development numerous times, mollusks and vertebrates have evolved rather similar camera eyes, and the giant anteater (a mammal) and the spiny anteater (a marsupial) have evolved similar digging and feeding specializations.

The evolution of such structures can be rapid. For example, the genus of tiny fish called *Poeciliopsis* evolved placental development in less than 750,000 years. The case of the octopus and the human camera eye has been looked into, and the lessons are clear. Underneath the gross anatomical similarities, illustrated in Figure 38, are many differences. The eye derives from different tissues by different developmental means. Although both structures use the same pigment (rhodopsin) for photoreception, and both send electrical signals to the brain, we now know that the intervening circuitry is completely different. Nonetheless, both have drawn on various cellular and developmental processes and components of the toolbox common to bilateral animals, using different tools in a different order. That the phototransduction circuits are completely different (involving components that are different but common to both organisms) is a testimonial to the power of conserved processes—they can be organized by different means to a similar end. In convergence, similar outcomes are evolved in different

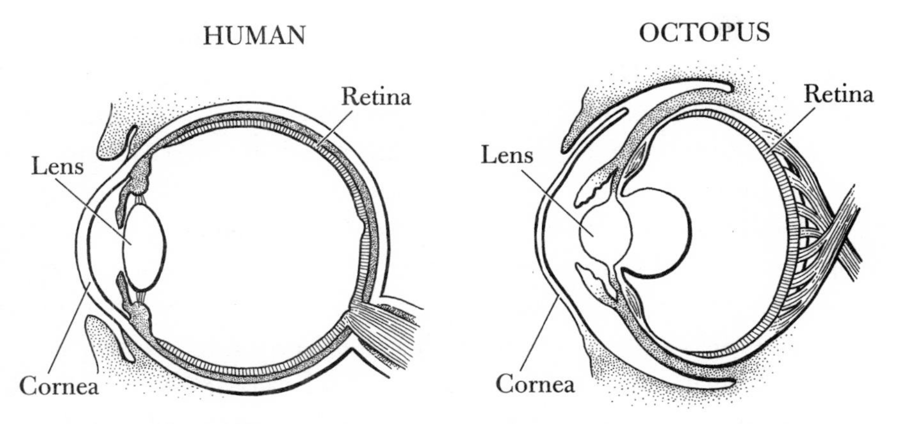

Figure 38 The superficial convergence of the eye of the octopus and the human. Both are camera eyes, but they differ in anatomical details, develop by entirely different means and communicate signals differently.

ways, making use of exploratory processes, modularity, flexibility, and weak linkage. Anatomical convergence at the level of these processes is no different than anatomical diversification.[12]

Evidence of facilitated variation could also come from measures of the genetic variation stored in wild populations. The high level of such variation is well known for numerous animals. In marine organisms such as sea squirts and sea urchins, 1–5 percent of base positions in the DNA differ between individuals of a species, a very high level revealed by genome sequencing. Greater somatic adaptability may correlate with larger amounts of genetic variation in a population. We have emphasized the robustness and adaptability of the core conserved processes. Just as these pervasive properties can buffer against external environmental changes, they can buffer against changes within, caused by mutational change. The greater the robustness and adaptability of core processes, the greater the tolerated random genetic variation, which is not eliminated by lethal effects and lessened reproduction.[13] The correlation of somatic adaptability and genetic variation has yet to be proved, and the nature of the processes that increase the amount of genetic variation is still not identified. The efforts by Lindquist and colleagues to characterize buffering mechanisms on a molecular level (for instance, the Hsp90 chaperone), and the efforts of geneticists to identify factors that affect the phenotypic consequences of mutations, will contribute to our understanding of how lethal phenotypic variation is suppressed.[14]

We can expect that the increased study of evolution and development, coupled with genomic analysis, will provide more and more examples of the use of facilitated variation in specific evolutionary events. In select genetic systems such as stickleback fish or in specific developmental systems such as bird beaks, we can expect more and more experimental tests of how conserved processes are deployed and more and more evidence of the preexisting poised processes that are evoked by small mutational changes. Facilitated variation will assuredly be exposed to further tests and refinements in the near future.

Life Without Facilitated Variation?

If we take the reverse tack, can evolution be imagined without facilitated variation? What capacity to evolve would a hypothetical organism have if it did *not* have facilitated variation? If animals did not use and reuse conserved processes, they would, we think, have to evolve by way of total novelty—completely new components, processes, development, and functions for each new trait. Under these circumstances the demands for "creative mutation" would be extremely high, and the generation of variation might draw on everything in the phenotype and genotype.

During the last half-billion years, the anatomical and physiological evolution of multicellular animals has not depended on total novelty, according to what we can ascertain from the fossil record and from comparisons of existing organisms. Even granting facilitated variation a big role in this period, from the Precambrian to the present, we have to admit that the conserved processes themselves had to evolve at some prior time, as did their special properties. Facilitated variation assumes the availability of these processes. The evolution of these processes and properties would seem to be the primary events of evolution, requiring high novelty. As noted in Chapter 3, the unique and episodic appearance of these processes with the emergence of eukaryotic cells, multicellular animals, and perhaps the first prokaryotic cells, may attest to the rarity of their invention. Once the conserved processes were available, though, the possibility of variation by regulatory shuffling and gating of these processes was unleashed, and shuffling and gating were much simpler than inventing the processes.

The main accomplishment of the theory of facilitated variation is to see the organism as playing a central part in determining the nature and degree of variation, thus giving selection more abundant viable variation on which to act. Several evolutionary biologists have argued that the organism should strongly influence the amount and nature of phenotypic variation resulting from genetic change. Lacking knowledge of the molecular mechanisms of embryonic development that underlie these developmental hypotheses, scientists could make little progress until the end of the twentieth century.

It is the capacity of the core processes to support variation that we see as the main factor in generating phenotypic variation and in minimizing the lethality of phenotypic variation. It is the nature of these processes, which are poised to generate physiological variation within the organism, that allows genetic variation to be so effective in generating phenotypic variation on which selection acts. Facilitated variation is a theory driven much more by mutation and genetic variation than the Baldwin-Schmalhausen ideas, which rely mostly on the environment to evoke change. We think that the organism is so constituted that its own random genetic variation can evoke complex phenotypic change. However, it is the extraordinary power of the conserved core processes that is most responsible for the copious amount of phenotypic variation in response to mutation.

EIGHT

Is Life Plausible?

We now take a perspective outside the theory of facilitated variation to measure its explanatory power. Having asserted that facilitated variation is consistent with neo-Darwinian theory, we ask, To what extent is evolutionary theory now a complete theory? To incorporate facilitated variation fully into the neo-Darwinian theory, we have to understand why it should be selected and maintained in populations. Many evolutionary biologists believe that any process that provides a future benefit must have an immediate benefit in each generation, or it will be lost. Whether or not such a stringent view is required, we still must evaluate the conditions under which facilitated variation can be selected. We have hinted at arguments for the selection of facilitated variation in passing; here we bring them together and evaluate them.

In asking where, after facilitated variation, our understanding of evolution is still incomplete, we confront the origin of the conserved core processes. Although we know little of the rare bursts of novelty that accompanied the evolution of eukaryotic cells or of multicellularity, we can see traces of their origins in earlier cellular processes and DNA sequences. The few hints we have suggest that when new core processes arose, there was an extreme modification of protein components, unlike the regulatory variation we encountered in facilitated variation. In trying to apply the lessons of facilitated variation to evolutionary history at the earliest stages, we revisit the episodic evolution of core processes and trace a scenario of how they arose and became central to evolutionary change.

Stepping outside science itself to view the possible impact of [illegible] variation, we consider some aspects of the influence of [illegible] in general society. Our goal is not to analyze this [illegible] far too soon to do so, but to indicate in two cases [illegible] should expect an impact. In the past, neo-Darwinian [illegible]as been a powerful metaphor for other areas of science and [h]uman engineering and design. Although aphorisms like "survival [o]f the fittest" have been used merely to extol winners and vilify losers, Darwin's concept of variation and selection has nevertheless been a useful inspiration for many ideas and designs, including some scientific mechanisms far removed from evolution. We will indicate how the inclusion of facilitated variation in the theory of evolution may have some application to engineering and institutional design at the beginning of the twenty-first century.

Among its worldly impacts is the role of evolutionary theory in the politicization of the teaching of biology in this country's public schools. Rather than being abstruse scientific issues, evolution and biology are emotional topics that repeatedly resurface in American politics. In the search for ways around the U.S. Constitution to question evolution and to force its removal from the public-school curriculum, opponents of evolution have retreated to a peculiar corner of criticism that questions the origin of novel traits. However, some of their favorite claims convert to strong arguments for evolution when one includes lessons from facilitated variation. Therefore, an understanding of facilitated variation could provide effective support for evolutionary theory in the social, religious, and political battles ahead.

Facilitated Variation and Evolutionary Theory

We begin this chapter with a return to Darwin's theory and an examination of how facilitated variation affects it. What has not been possible to estimate in evolutionary theory and has not even been a prominent part of evolutionary discussion, is the exact nature of the dependence of phenotypic variation on genetic variation. How does genetic variation generate phenotypic variation? Can it generate enough, and of the right type? Do all properties of the phenotype change? If change is biased, what is the bias and how does it arise? These are questions answerable only in the new millennium, after the establishment of genome sequences and the broad understanding of cellular and developmental biology.

One goal of this book is to understand what conservation says about diversification, especially in animals. Conservation, we believe, facilitates diversification and reveals the nature and plausibility of evolution as a unified theory to explain life's vast diversity. Facilitated variation definitely implies a biased output of phenotypic variation by an organism (at least of the anatomical and physiological kind), even though the initial input of mutation over the entire genome is random. This bias is inevitable, because variation is based on reuse of the existing phenotype in new ways and hence starts with a given structure, a given bias.

For animals from the Cambrian to the present, one can say that variation has mostly concerned changes of anatomy and physiology, based on the deployment of conserved developmental processes in different combinations and amounts. Even though the conserved processes and body plans at hand have sufficed for a great range of morphologies over the past 600 million years, conservation has biased the output, the ways that organisms can express their diversity in their offspring. The realm of raw possibilities, we suggest, would have been much larger had random mutation really been able to cause random unconstrained phenotypic variation, that is, de novo originations unrelated to what has already evolved.

While pure random phenotypic variation might have produced far greater diversity and less consistency between organisms, it might also have been more lethal and detrimental to reproductive success. Random mutation, in the view of facilitated variation, selects or channels phenotypic variations via regulatory changes, rather than creating them. Evolution has been compared to a biased random walk because of the bias introduced by selection, but we say phenotypic variation is itself biased. Rather than staggering like a drunken sailor, evolution marches along a myriad of paved pathways, changing direction without instruction, but taking large, forceful steps and avoiding many lethal obstacles.

With facilitated variation, the tripartite Darwinian theory consisting of genetic variation, phenotypic variation, and selection becomes much more complete. As we learn still more about cellular processes and their ranges of adaptability, facilitated variation will begin to constitute a theory of what is possible in biology, indicating preferred and forbidden paths of phenotypic variation. But a theory of the possible may still fall short of predicting the actual, since uncertainties in the environment and the organism, fluctuations or noise in the organism's development, and the randomness of mutation, recombination, and reassortment will impinge strongly on the actual path of evolution.

Despite the inclusion of facilitated variation, natural selection stands as a robust and important part of Darwin's theory. Although facilitated variation biases the amount and nature of variation available for selection, the variation possible from the recombining of adaptable core processes is very large. Hence, variation is abundant, and natural selection still molds what is presented to it. Facilitated variation is not like orthogenesis, a theory championed by the eccentric American paleontologist Henry Osborn (1857–1935), which imbues the organism with an internal preset course of evolution, a program of variations unfolding over time. Natural selection remains a major part of the explanation of how organisms have evolved characters so well adapted to the environment.

Darwin's original view and the neo-Darwinian reinterpretation, which assumed copious amounts of very small variation in all directions, relied solely on natural selection as the determinant of exquisite evolved adaptations. The reason why variation in the workings of a machine like Paley's watch is almost always catastrophic, whereas such variation in an organism is not, speaks to the difference in how the two are constructed. In particular, it is a comment on the rigidity of the components of the watch and the adaptable nature of the organism's core processes.

Although it may come as little surprise that the organism is adaptable and that adaptability plays a role in evolution, what we have added from the trove of modern research findings is the chemical nature of that adaptability and the history of how it originated and how it has been used. These insights allow us to discuss more deeply the role of adaptability in evolution. The bias introduced by facilitated variation accelerates the process of natural selection by giving it more viable variation of a type likely to be appropriate to the selective conditions than it would have been if variation occurred in all directions. Thus bones, beaks, and the physiology of the heart and nervous system are modified in directions more likely to generate viable animals. This perspective moves natural selection from a theory of small changes to one that can explain the origin of significant novelty in evolution over short periods. Although we are left with a more complete theory of evolutionary change, we still have the question of how facilitated variation arose in the first place and why it has been maintained.

Selection for Evolvability

"Natural selection will favor traits that enhance the possibility of further evolution, and so will reveal evolvability to be the greatest adaptation of all." David DePew and Bruce Weber thus trumpeted the transcendent importance of evolvability. Was evolvability selected? It is difficult in these terms to devise airtight arguments. The surely *un*convincing one is that evolvability has been selected for its future benefits, as implied in the quotation just above.[1]

What, instead, might be the selective advantage of evolvability? It is one matter to find convincing arguments for why a mechanism was initially selected, and it is another to find arguments for why it has been maintained in the population thereafter. In the former case, we should look for arguments that address current advantages, whether they are directly related to evolvability or are by-products of some other process. For the latter, we can easily revert to the conservation argument, namely, that once a variation-generating mechanism is present, it is maintained simply because it is repeatedly reselected with new traits undergoing selection. We have to be careful in arguing that if some organisms have evolvability mechanisms and others do not, the former will evolve better and eventually displace the latter. Although the argument may be valid, it can be circular. Furthermore, we are raising arguments not just for evolvability in general but for facilitated variation in particular.

Several arguments, taken together, show that the means of facilitating variation should be under positive selection. All concern why conserved components and processes have been selected to be the way they are—that is, robust, flexible, versatile, adaptive, capable of regulatory weak linkage, exploratory, modular, and prone to compartmental usage. Selection can be for current use, but the benefits can be long term as well as short term.

Our point is that conserved core processes respond to genetic variation or environmental variation by producing their special type of phenotypic change. They are particularly effective at accommodating to these inputs and reducing their lethality. These responses may be temporary and reversible, in which case we call them somatic or physiological adaptations, or they may be stabilized by new mutations or genetic reassortment, in which case they may become heritable and evolutionary adaptations.

Four general statements summarize how facilitated variation either directly or indirectly confers current benefits on the individual or future benefits on the group level.

1. The processes that generate facilitated variation are selected for their contribution to *effective development and physiology*. Robust, adaptable processes are best for the development and physiology of complex multicellular organisms, which proceed under inevitably variable internal and external conditions. Processes with these characteristics are initially selected *directly* for their function in the organism. Facilitated variation is then simply a by-product made available by these characteristics. By-product status does not belittle facilitated variation; in fact, it is inspiring that the

conserved processes are suited for both life in the short run of each generation and life in the long run of evolution.

2. The processes that generate facilitated variation are selected for their contribution to *descent with modification*. Processes that are prone to multiple use will contribute to facilitated variation. Similarly, processes that generate multiple uses easily can contribute to the physiological adaptability of organisms. These two selective advantages work in the same direction. The selection for multiple use of a process would be selection for its increased adaptability and robustness, that is, for its special properties. Processes in the cell that are capable of integrating more complex inputs such as environmental conditions, or more complex outputs such as expression of multiple genes, will increase the robustness of an organism and can be directly selected for the benefit they provide. Yet selection of processes that themselves have the capacity to function in multiple ways, as through weak linkage, increases the likelihood that new processes can be cobbled together in evolution. Such processes would therefore facilitate nonlethal variation in response to a given amount of genetic variation.
3. On a population level, facilitated variation contributes to an increase in *genotypic variation*. The properties of robustness and flexibility in the conserved core processes buffer against lethality and impaired reproduction. As a result, more genetic variation is kept in the population as viable variation, an argument first propounded by Schmalhausen. Such genetic variation is then available in a form that can be expressed phenotypically under stressful conditions, as was observed in Waddington's and Lindquist's heat-shock experiments to evoke aberrant anatomies. In the facilitated variation theory, most of this variation serves regulatory purposes, stabilizing new combinations of old processes and selecting different parts of the adaptive range of their outputs. This storage of variation is perhaps no

longer just an individual selective effect, but a population effect that might favor the persistence of lineages.

4. Facilitated variation may have been coselected on a group level during *evolutionary radiations*. The capacity for generating increased phenotypic variation may facilitate radiations of organisms into new or emptied niches. Radiations of large groups have happened several times: the radiation of insects on land and in the air; the radiation of vertebrates first in the ocean, later on the land, and still later in the air; the radiation of mammals after the dinosaur extinctions; the radiation of one species of fish, the cichlids, into newly formed lakes; and the radiation of one or a few bird species on volcanic islands such as the Galápagos and Hawaii. Richard Dawkins wrote: "Just as some organisms are good at flying or swimming, so may some be good at evolving. In particular, certain kinds of embryology may be predisposed to spawn rich evolutionary radiations." Facilitated variation may be of prime importance in periods of radiation of animal groups, when interspecies competition is less an issue than preemption of niches. Under these conditions, the capacity to generate variation seems particularly important, and increases in the adaptive and robust behavior of core processes might occur under selection.

If we assume that selection for facilitated variation is largely for the reasons stated above, then it is reasonable that different lineages would differ in their capacity to generate phenotypic variation. If superimposed on this variation are repeated radiations and extinctions, a further selection for the capacity to vary would take place, and those lineages richer in facilitated variation and hence richer in the capacity to radiate would be preserved.[2]

In the overall argument for evolvability by facilitated variation, we need to remember that genetic variation is always needed for heritable new combinatorial use of conserved processes and for heritable new use of a part of the adaptive range of a process. The mutation need not occur at the time of the selection; it may have been in the population for a long time. The organism as a massive system of conserved processes seems primed to respond to random mutation by way of regulatory changes, to give new outputs of the conserved processes. Although an input of genetic variation is needed, the output of phenotypic variation, as a response, seems rather like somatic adaptability, appropriate to the environment and solving some environmental problem. It was this adaptability that confused Lamarck and even Darwin.

In our view, the capacity for facilitating variation has itself evolved as the core processes of organisms have accumulated more adaptive and robust behaviors. Evolution does not proceed on random generation of dysfunctional phenotypes, which almost always result in lethality and only by accident give rise to an advantageous trait. Lethality is most an issue when genes are mutated that encode components of the conserved processes. These mutations are eliminated by selection in each generation (and many are probably eliminated in the germ line). Exempting those, the population accumulates genetic variation because of the robustness of physiologically adaptable processes, and the individual generates phenotypic variation in response to genetic change or environmental change that is predisposed to be less lethal.

In summary, we believe that evolvability—the capacity for organisms to evolve—is a real phenomenon. We believe that facilitated variation explains the variation side of evolvability, through the reuse of a limited set of conserved processes in new combinations and in different parts of their adaptive ranges due to genetic modulations of nonconserved regulatory components. In our view, an environmental stimulus may initially provoke the new phenotype in some cases, but eventually genetic change of a regulatory component is needed if the phenotypic change is to be heritable. In other cases, genetic variation may itself provoke a new phenotype.

Facilitated variation has arisen and increased by selection, we say. Since it facilitates the generation of innumerable complex, selectable,

heritable traits with only a small investment of random genetic variation, it is indeed the greatest adaptation of all, at least for animals since the Cambrian. On the side of generating phenotypic variation, we

believe the organism indeed participates in its own evolution, and does so with a bias related to its long history of variation and selection. Coupled with our already advanced understanding of natural selection and heredity, facilitated variation completes the broad outlines of the general processes of evolution, particularly for metazoan diversity.

The Origins of Core Processes

The theory of facilitated variation opens up a new set of questions about the origins of the conserved core processes that, as we have argued, facilitate the generation of all of kinds of anatomical, physiological, and behavioral diversity. There is really no alternative but to think that new core processes, such as those that first arose in eukaryotic cells, were cobbled together from the existing processes in prokaryotic cells. The transformations from prokaryotes to eukaryotes or from single-celled to multicellular organisms are profound, and evidence is sparse. However, as methods for identifying weakly related DNA sequences have improved, and as more organisms have been sequenced, we can glean hints about these major transitions.

Core processes may have emerged together as a suite, for we know of no organism today that lacks any part of the suite. For example, there are no eukaryotes without mitochondria in their ancestry. (Some organisms that once had energy-producing mitochondria no longer have them, but all show traces of once having had them.) Thus, in extant eukaryotes we meet the whole span of eukaryotic processes whenever we meet a eukaryotic cell: mitotic separation of chromosomes, transport of materials on the cytoskeleton, and compartmentation of cell functions by membrane boundaries.

Until the 1990s, the cytoskeleton of eukaryotic cells had no known precursor in prokaryotes. Since prokaryotes lack a mitotic spindle and cannot crawl along surfaces, no one expected them to contain proteins like tubulin and actin (the proteins that make up microtubules and microfilaments in eukaryotes). However, distant protein relatives of both tubulin and actin have now been discovered in prokaryotes—thus, there are clear intimations of a cytoskeleton prior to eukaryotes.

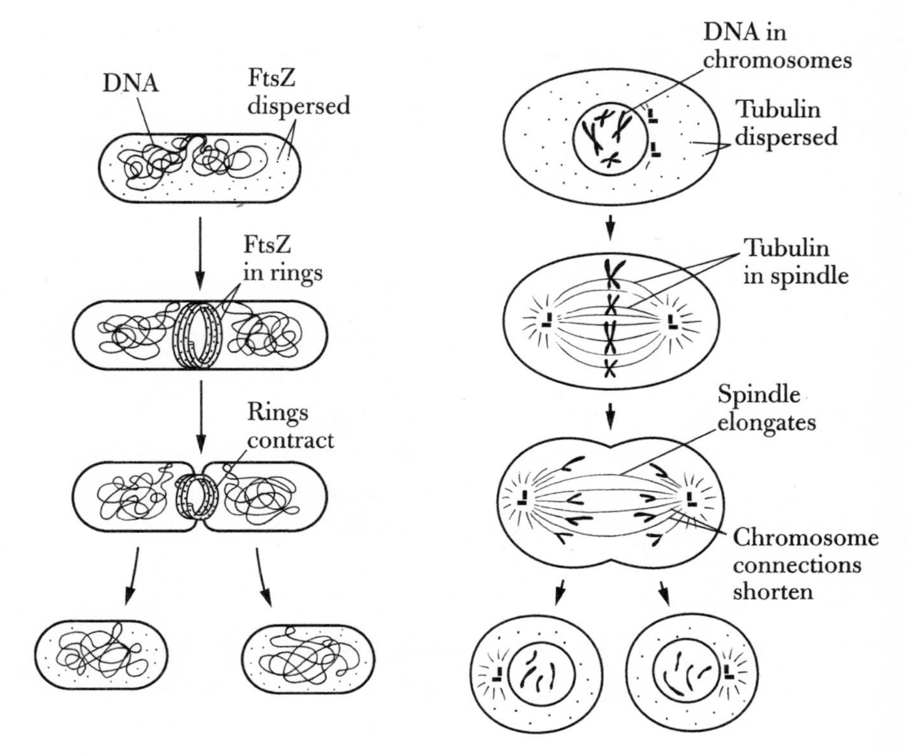

Figure 39 The prokaryotic origins of the cytoskeleton. The tubulin protein of animals (*right*) may have evolved from the tubulin-like FtsZ protein of bacteria (*left*). Both proteins have the same shape, but they form long filaments of different structure; both participate in cell division, but many of the details differ greatly.

Although the bacterial proteins show almost no sequence identity to their eukaryotic relatives, they share virtually the same three-dimensional structures. Bacterial tubulin (called FtsZ), like its eukaryotic relative, forms a dynamic structure involved in cell division, but that structure is completely different from the mitotic spindle, as shown in Figure 39. Furthermore, the function of these proteins in eukaryotes is much more diverse than in bacteria.[3]

The harder we look, the more evidence we find that the typical eukaryotic functions have distantly related counterparts in prokaryotes, supporting the idea they must have been present in a common ancestor when eukaryotes split off. In all cases, however, major changes have occurred in the sequence of the proteins and in their use during the evolution of the new processes. For example, chromosomes are a hallmark of eukaryotic cells, seemingly unique to them, but relatives of some of the proteins involved in the segregation of chromosomes at mitosis are also found in bacteria and play a role in segregating DNA. Membrane-bounded internal compartments are a characteristic of eukaryotic cells. Proteins are moved through these compartments to the outside of the cell, which might seem an ability unique to eukaryotes. Yet even without internal membranes, bacteria secrete proteins through their surface membrane by a similar process that utilizes proteins related to those of eukaryotes.[4]

These cases suggest that the great innovations of core processes were not magical moments of creation but periods of extensive modification of both protein structure and function. The changes are not achieved by facilitated variation of the regulatory kind we have described throughout this book. Instead, during great waves of innovation, preexisting components of prokaryotes changed their protein structure and function in fundamental ways to generate the components of new core processes of the eukaryotic cell. These changes are clear when we compare the eukaryotic tubulin gene to its prokaryotic relative (FtsZ). Tubulin is a highly conserved protein in eukaryotes, and its prokaryotic homologue is equally conserved in bacteria. Yet tubulin and its bacterial distant cousin differ so much in sequence that they are virtually unrecognizable as relatives. These proteins are strongly conserved within their large groups (eukaryotes and prokaryotes), but the extensive sequence variation between the prokaryotic and eukaryotic tubulin cannot be explained by the many years that separate them; there are just too many changes. Instead, a period of rapid remodeling of the tubulin precursor is likely as eukaryotic cells arose in evolution; thereafter, tubulin changed very little.[5]

Further intimations of true novelty come from the time when multicellular eukaryotes including animals first arose a billion years ago. Animals produce a variety of proteins not found as such in single-celled eukaryotes. The protein novelty of this episode is mostly of two kinds. Many new proteins of large size were produced as new combinations of small, functional proteins similar to those found in single-celled eukaryotes. Large combinations of pieces are novel to animals. As discussed earlier, this kind of protein evolution has been undeniably facilitated by the exon-intron structure of eukaryotic genes and by the capacity of eukaryotic cells to "splice" RNA transcripts. (The splicing capacity functions continuously in the individual in the production of messenger RNAs in cells.) The new proteins participate in multicellular functions, such as adherence to the extracellular matrix, cell-cell communication, and tissue reorganization through formation of intercellular junctions.[6]

The second kind of novelty arises from the duplication of old genes, followed by the divergence of protein coding sequences to give related but distinctive functions. For example, protein kinases have diverged from one or a few kinds in single-celled eukaryotes to over a thousand kinds in vertebrates, differing in details of their target specificity. Also, transcription factors related to Hox genes have diverged from one or a few ancestral proteins into several hundred kinds in vertebrates and insects, differing in their DNA-binding specificity and their interactions with other proteins. As in the prokaryote-eukaryote transition, this transition to multicellularity drew in large part on preexisting components for the generation of protein novelty.

The most obscure origination of a core process is the creation of the first prokaryotic cell. The novelty and complexity of the cell is so far beyond anything inanimate in the world of today that we are left baffled by how it was achieved. Unlike the later revolutions, no prior core processes and components were available for modification to make the first cell, or at least none has survived. Lacking any example of an organism that might have diverged before the common ancestor of bacteria, we can do little more than speculate. All we know is that today there is only a single lineage of life (that is, a single DNA-RNA-protein machine and a single metabolism). Perhaps, for all we know, life originated only once.

After that first (and most significant) innovation, the origin of life, the subsequent several waves of innovation are increasingly supported and reinforced by observations and experiments coming from several directions. We may not have a full account of the origin of each of the subsequent core processes, but we have enough shards of evidence to see that they originated from existing proteins and then underwent considerable modification of structure. By contrast, evolution since the Cambrian is supported by irrefutable molecular evidence and a compelling fossil record.

A Moving Front

Having an established theory of phenotypic variation based on facilitated variation and some understanding of how core processes arose, we are ready to contemplate the history of evolutionary change. Does it reflect a history of continuous branching from a common origin, as Darwin saw it? Is there a clear correspondence between periods of evolutionary diversifications seen in the fossil record and the underlying invention of conserved core processes? We can now revisit the relationship between morphological and physiological divergence and the invention and implementation of the conserved core processes not only with the aim of providing a description but with the hope of some explanation.

From the perspective of facilitated variation, life on this planet is divided into several epochs of cellular innovation, and these epochs do not correspond to known epochs of transforming geological events. The conserved core processes appear to have been added in stages—in several relatively short episodes separated by long intervals when no major core processes were added. The core processes were maintained from then on.

These bursts of addition of core processes prompt speculation about the evolution and implementation of facilitated variation. We suggest the following scenario: first, the generation of novel processes, which we described above; second, the acquisition of robust, adaptive properties by each process; and third, the rampant regulatory usage of processes in a way that facilities the generation of phenotypic variation.

New Core Processes

The processes and components went through an initial phase of evolution, whereby their overt function was established. As we saw in the cytoskeletal proteins, old genes of prokaryotes were radically retooled for different but not totally unrelated purposes in eukaryotes. Here novelty was really called for. Protein coding sequences of genes, rather than regulatory sequences, changed greatly. The difficulty of generating new suites of conserved mechanisms may be reflected by the vast gaps between bursts of additions—from the first prokaryotic life to eukaryotes, then to metazoa, then to the body plans.

Increasing Robustness and Adaptability

The new processes probably had limited function, in that the properties that would allow them to integrate with existing functions were not yet established. Perhaps the cytoskeleton served only for mitosis or for the transport of only one kind of membrane-bounded vesicle. In multicellular development, perhaps cell signaling was limited to a few pathways, and those to only a few cell types. During integration, the now-functioning processes and components would have undergone modification toward robustness, flexibility, compartmentation, exploratory behavior, and capacity for weak linkage. They could function well despite variable conditions both outside and inside. They could work in combination with other processes, and they were easily connected to other processes. Their capacity to buffer environmental and genetic variation increased. More and more components of the process interacted among themselves as integration occured, and they perhaps become constrained against further mutational changes of the coding sequences. The period of conservation began.

The processes, though constrained to change within themselves, were now deconstraining in their regulated interaction with other processes, by their robustness, adaptability, and capacity for weak linkage. They were capable of diverse outcomes, as seen in those with exploratory behavior. The capacity for variation had increased, in the sense of the readiness of processes for use in different combinations and amounts to different ends.

Rampant Regulatory Usage

In this final phase, observable today, facilitated variation plays out. The internally constrained processes, with their adaptive capacities for weak linkage, exploratory behavior, compartmentation, robustness, and flexibility, by various regulatory means are used in manifold combinations with other processes, and in different parts of their adaptive ranges. Evolvability increases, and phenotypic radiations occur. Conservation of core processes is strengthened as the processes and components are repeatedly reselected with each selected trait they participate in generating, each trait being a new combination.

Once a round of novelty begins, after some interval from the previous period of innovation, the three phases repeat. As noted before, the conserved processes of a particular period may set a boundary on the realm of possible variations that regulatory mutation can explore. The realm is presumably larger at each successive period, complexity building on complexity. It really exploded at the multicellular stages, because of the possibilities for gene expression in different cells at different times and places in the huge population of cells of the individual multicellular animal. Facilitated variation is seen to have taken take a giant step after the core processes of multicellularity were introduced, as it did at previous steps of core process evolution.

There is no indication that we have depleted the kind of phenotypic variation that can be extracted from regulatory control of the core processes we inherited 600 million years ago, when aspects of the body plan were originated. New mechanisms of facilitated variation have arisen since then; for example, imaginal discs and larval development in insects, and limb buds and neural crest cells in vertebrates. The development of the neocortex in mammals is highly adaptable, and that brain region has gradually taken over functions from other parts of the reptilian brain. Small arenas of innovation exist with important ramifications; think of the mechanisms of dentition in various vertebrates, and the capacity to change tooth structure rapidly with the concomitant increased opportunities for feeding.

There is every reason to believe that facilitated variation itself has evolved in specific ways via introduction of the higher-order core processes involved with embryonic development. Yet even if the capacity for facilitated variation varies and widens its scope or improves its efficiency in generating phenotypic variation, we do not want to imply that these advances constitute progress toward a preset goal. Getting better at evolving is not the same as evolving for the better.

Other Kinds of Complexity

Bearing in mind the often tendentious application of Darwinian models to economics and politics, does facilitated variation have something useful to contribute to understanding complex social or political organizations, or elements of design in engineering, or computer science? We might expect that facilitated variation, which emphasizes the means to generate variation and diversity, could bring a different perspective to the examination of structures and institutions than did natural selection, which emphasized selection, survival, and reproductive success. The latter, promulgated in the early part of the twentieth century as Social Darwinism, has often been used as an expedient means to justify the current order of things as naturally destined. We might be on firmer ground to consider the value of variation and diversity in their own right and leave the question of selection and success to a more nuanced analysis.

The caveats against going in this direction are strong. We run the risk of trivializing the fundamental differences and exaggerating superficial similarities, some of which may arise merely from overlapping terminology. We also risk succumbing to the polemical temptation to argue that a certain model of a social system should be adopted because it is used in the natural world. Let us proceed with caution.

On the positive side, models as metaphors or points of inspiration sometimes can be moved profitably from field to field. Models from engineering and physics have been used in biology to good effect, most recently in understanding the control of systems. The use of biology as a model in engineering has been prominent in genetic algorithms in computer science, though their applicability so far has been limited. That a comparison of different complex systems might yield new insights for both is the promise that urges a tentative and highly qualified exploration of these areas.[7]

We have described organisms as complex systems of hundreds of conserved core processes, all having adaptive ranges of operation, organized in different combinations in the many compartments of the body plan. In organisms, the combinations all operate in parallel and occur in several other dimensions of compartmentation as well.

John Doyle, a mathematician who has studied complex systems in engineering, argues that a new engineering, arising from computer science and communication science, shares significant similarities with biological systems. Both computer science engineering and biological systems appear to have features of modularity, robustness, and rules for interaction that are general and extendable. They are susceptible to catastrophic failures, like our electrical grid. He calls attention to the "spiraling complexity" of engineered and biological systems, referring to the ever more extensive regulatory circuitry added to the increasing number of functional components in order to achieve robustness. The utility of control theory and engineering as applied to biology, and the similarity of some of the concepts in biology to the new engineering of hypercomplex systems, suggest that overriding themes may unite various systems productively. Both biology and engineering have undergone a transition in the size and complexity of the subjects they contemplate; by comparing these systems, we may be able to derive new rules of design.[8]

We have been considering the question of how the particular structure and properties of a biological system, the organism's phenotype, increase its ability to generate heritable variation. We have defined organisms as frugal in their design; they use a limited number of components in combinatorial ways to convert a small number of mutations into phenotypic novelty. We have found that many of the systems are profligate in their use of resources, regularly creating excess cells or axons or microtubule arrays, all in the course of generating a specific architecture. These exploratory systems are very robust, yet their robustness is offset by the wasted energy of creating many structures that are of no immediate use to the organism. We have argued that such properties reduce costly or potentially lethal variation, and that they therefore have a significant role in facilitating evolution. Variation in biology is inhomogeneous, restricted primarily to regulatory components that determine time, place, amount, circumstance, and orientation, as well as the part of the adaptive range to be used. Special features of the core processes make them robust and adaptable in the short and long term: weak regulatory linkage, exploration, and modularity or compartmentalization. These processes have not changed over time, but they deconstrain regulatory change around them. Somewhat to our surprise, processes that are used for short-term somatic adaptation are often modified in a more permanent way for heritable long-term evolutionary adaptation. Finally, the conserved core processes are robust to damage, and as a result the organism tolerates genetic variation in the population.

Is this the kind of system we would like running our car, our school system, our company, our government? We may not want to incorporate all (or any) of these features into any institution or plan. After all, institutions are generally constructed from the top with premeditation, and it is usually assumed that directed processes are the most efficient. However, any system has features of local autonomy, self-organization, and flexibility, so we might expect some of those to carry over from biology to more hierarchical systems. Also, if we really are on a course of spiraling complexity, it might be hard to design a centrally controlled system—and if one could, it might be too vulnerable. Perhaps designers of future computers or institutions will intentionally borrow features from facilitated variation.

Human institutions, for all their differences from other biological systems, may have some of the same requirements as organisms. (The metaphorical comparison of organizations to organisms is a very old one, and we recognize that we could go down the same sterile path taken by others in the past.) To the extent that cultural evolution is analogous to biological evolution, the same need remains to preserve core functions while using them differently. The lesson from facilitated variation is that great care should be taken in generating the core processes and their properties, for they will be conserved and will determine how much deconstraint surrounds them. Weak linkage, as a property, implies a simple and uniform capacity to interact and to change the interactions—common examples are wall jacks for computers and telephones. (As we lug around various electrical adapters in our travels, we are all aware of times when we have overlooked the capacity for weak linkage.) Exploratory behavior may be important if change is not to be restricted to combinations of existing behaviors and protocols. Modularity in biology is a common and stable property; for example, the compartment maps of the body plans of phyla. The spatial divisions of the map are quite arbitrary (although they may have represented anatomical function when they were first invented in the Precambrian). Still, they establish different noninterfering locales. They function well as long as they possess great flexibility in altering the rules for local function. Why biology conserves the spatial modules and varies the rules within each module, rather than vice versa, is perhaps worthy of thought and investigation in areas outside biology.

The fact that robustness and adaptability yield individual diversity in the population, and that diversity is a useful feature, has another resonance in complex human societies. As the robustness of biological systems increases, other things being equal, diversity increases. The amount of diversity is one measure (after corrections for history and population size) of the effectiveness of the robust adaptable mechanisms that facilitate variation.

Finally, the paths of major innovation in human society may parallel the process of major innovations in the core processes in biology. New core technologies may be adapted through a sequence of novelty generation, integration into existing protocols and products, followed by rampant use similar to the rampant regulatory variation in the moving front of innovation that we have just described.

At the very least, an analysis of evolvability by facilitated variation evokes different metaphors than does Social Darwinism, which stressed selective conditions, not variation. History is not just a product of selection, determined by the external environment or competition; it is also about the deep structure and history of societies. It includes their organizations, their capacity to adapt, their capacity to innovate, perhaps even their capacity to harbor cryptic variation and diversity.

Perhaps the most important lesson of this analysis may be that the generation of phenotypic variation from genetic variation cannot be taken for granted in complex organisms: it does not reflect random breakdown of the system, but rather a selected design mode of the conserved core processes. That may also true in complex organizations. To achieve variation (especially nonlethal variation) requires that the rules and properties of the elements be designed in a special way. Although these rules and properties may constrain change within certain elements, at the same time they deconstrain changes of overall short-term and long-term behavior. It may be useful to look for such features in other systems.

Creationism and Intelligent Design

Though modern scientists may have questioned the completeness of the theory of evolution, few believed that the fundamental principles of variation and selection would not in the end explain the diversity of life. Certain groups, however, particularly active in the United States, have exaggerated and fabricated weaknesses in evolution theory in order to discredit it. From its beginning, the theory of evolution has caused problems for some traditional religious groups. By depicting human beings as derived from simpler animals, evolutionary theory not only undermined the biblical account of creation, but also seemed to debase human beings by suggesting that they were not of divine origin.

Not all religious groups chose evolution as a battleground. As early as 1909, the *Catholic Encyclopedia* wrote: "This conception is in agreement with the Christian view of the universe. God is the Creator of heaven and earth. If God produced the universe by a single creative act of His will, then its natural development by laws implanted in it by the Creator is to the greater glory of His Divine power and wisdom."[9] We ask here whether facilitated variation, as an explanation for the generation of novelty, can defuse some of the controversy that continues to rage over the teaching of evolution in the American public schools.

Consonant with the strong constitutional barrier dividing the state and religion, the U.S. Supreme Court in 1987 prohibited the teaching of the biblical story of creation as science in the public schools. This ruling stimulated a new effort by the opponents of evolution to find nonreligious critiques, to argue that contradictory evidence could now be presented as a kind of science along with evolutionary theory, without violating the separation of church and state. These secular theories, labeled intelligent design, argue against evolution not only on factual grounds, by finding controversies and ostensible flaws in evolutionary theory, but also on theoretical grounds, by attempting to show from first principles that evolution is impossible. A very small number of scientists (and almost no biologists) have shared in the skepticism.

From the viewpoint of facilitated variation, something is particularly intriguing about the polemical strategy of the proponents of intelligent design. Among the three pillars of evolution theory, their favorite target has been the origin of novelty. Although they question the fossil record, citing disagreements among scientists, proponents of intelligent design specifically deny the possibility of an origin of novelty in biology. Natural selection as a process gets less criticism, except as part of an argument that the improbable varieties were never there in the first place to select from. Modern genetic mechanisms are also rarely questioned. As phenotypic variation is the least understood of the theoretical underpinnings of evolution theory, it may not be surprising that it is currently the favorite target.

By arguing for an intelligent designer, creationists have sought a false completion of evolutionary theory, generating for the faithful a sense of satisfaction in what for them was an unexplainable system. In this book we have addressed this incompleteness in evolutionary theory by assembling scientific evidence for the causes of variation. These conclusions bear on the issue of intelligent design. Even if the promulgators of intelligent design are merely covertly advocating their own religious agenda and have no desire to hear alternatives based on modern molecular and developmental research, other more open-minded people may be influenced by arguments for the plausibility or implausibility of generating novelty in evolution. So, without trying to answer all the ostensibly secular attacks on evolutionary theory, let us weigh what we can add to the argument. The following are three of the more prominent cases for intelligent design, and the contrary view from the proponents of facilitated variation.

About a decade ago, a major book on intelligent design was *Darwin on Trial*, which discusses many issues relevant to facilitated variation. An example is Ernst Haeckel's famous 1874 drawings, which illustrated the similarity of various vertebrate embryos at the phylotypic stage. Various authors have commented that Haeckel, with artistic flourishes, overstated the anatomical similarity. The author of *Darwin on Trial* finds a much more serious error: "Although it is true that all vertebrates pass through an embryonic stage at which they resemble each other, in fact they develop to this stage very differently. Only by ignoring the early stages of development can one fit Darwin's theory to the facts of embryology." In other words, if animals are closely similar at a middle stage of development, they must have been at least as similar at earlier stages. The conclusion is, "If embryology is our best guide to genealogy, . . . vertebrates have multiple origins and did not inherit their similarities from a common ancestor."[10]

Objection! Today we recognize the assumption and conclusion as erroneous. A wealth of experiments and interpretations have converted the diverse anatomy of early vertebrate stages from a confounding paradox of evolution to one of its strongest arguments. The real anatomy of the conserved phylotypic stage of vertebrates is not simply the overt shapes and bumps that Haeckel may have embellished, but a highly conserved map of compartments of expressed selector genes, whose functions can be tested individually by genetic experiments.

An unbiased view of some of these embryos and the corresponding selector gene domains is shown in Figure 40, where the anatomical similarities are still obvious. The figure also shows the conserved selector gene compartments. On the matter of how the phylotypic stage can be conserved while the preceding stages are not, we raise the evidence of the regulatory deconstraint provided by the constrained compartment map. This deconstraint, we argue, has allowed considerable variation in the pathways (and anatomies) that emplace the conserved compartment plan, and has allowed the egg and embryo to evolve other pathways of development. These pathways operate while the compartment map develops, independent of it, and furnish the embryo with nutrients and protection. Not only vertebrates, but also arthropods and other phyla, must be judged by their conserved compartment plans and selector genes, not by the anatomy of their early stages of development, at which time special adaptations are as rife as in the adult.

Advocates of intelligent design have introduced the term *irreducible complexity*. It is meant, in principle, to contradict the theory of evolution by arguing that complex physiology is too improbable to have ever been assembled by chance. Michael Behe calls attention to "systems of several well-matched, interacting parts that contribute to basic functions, wherein the removal of any one of the parts causes the system to effectively cease functioning." He continues: "Now that the black box of vision has been opened, it is no longer enough for an evolutionary explanation of that power to consider only anatomical structures of whole eyes, as Darwin did . . . Each of the anatomical steps . . . actually involves staggeringly complex biochemical processes."[11]

We can sympathize with Behe, with Paley, and even with Darwin, that the origins of extremely complex structures are hardest to understand. Though the advocates of intelligent design invoke "irreducible complexity," they never ask about the nature of that complexity. Behe uses elaborate biochemical examples to intimidate us into believing

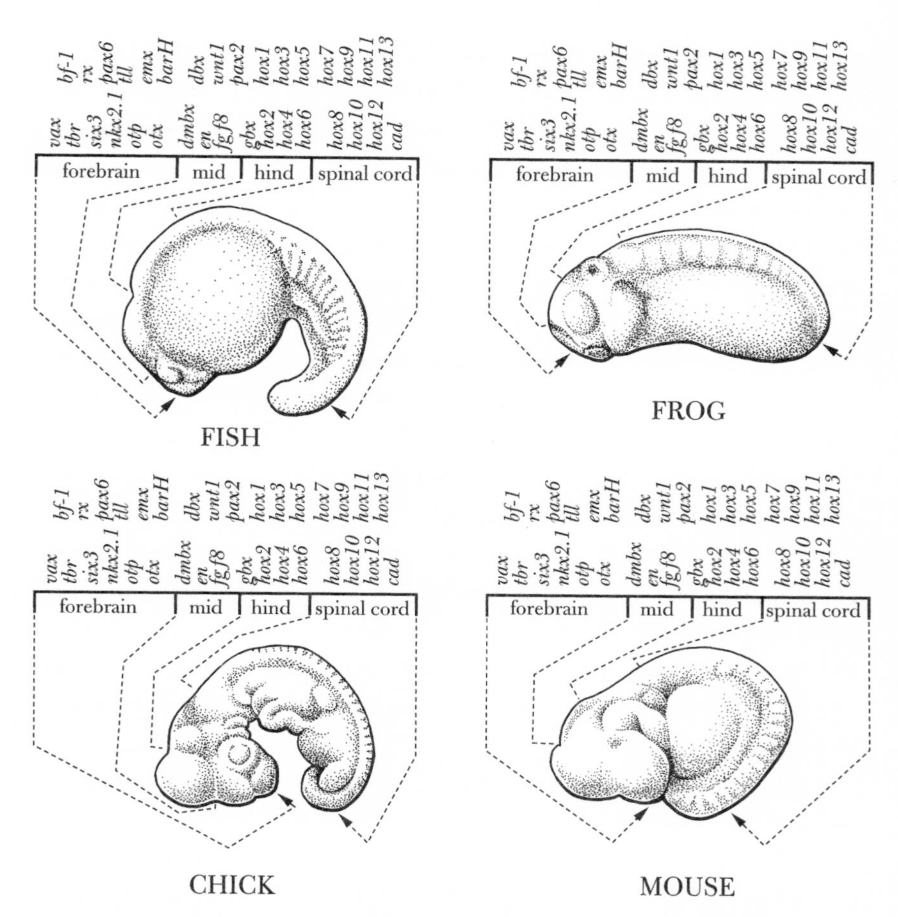

Figure 40 Haeckel reconsidered. Four embryos of different vertebrate classes are shown at their phylotypic stages, when they look most similar. Although their shapes are distinguishably different, they have the same map of compartments of selector gene expression (indicated by gene names above each drawing). The map has just formed at this stage. Extraembryonic tissues have been removed from the chick and mouse embryos.

that the complexity of living cells is beyond understanding. Yet today, understanding the nature of complexity is a major pursuit in science and the focus of much of this book.

In Behe's particular example, we know that the signaling pathway from the visual pigment (which itself is conserved from bacteria to humans) to the electrical channel in the cell that receives the light impulse in the retina is, in fact, a concatenation of conserved core processes common to eukaryotic cells. Furthermore, these processes all have a capacity for weak linkage so that they can be easily wired in different circuits. Ironically, one of the best examples of the capacity for weak linkage for rewiring is the eye. If Behe were to look at the biochemical pathway for vision in insects, he would find it almost completely different from that in vertebrates; but on delving deeper, he would see something more remarkable than two unrelated complex examples. Though the wiring in the insect and the wiring in the vertebrate are completely different, the components used in the two visual systems are again taken from a shared dowry of conserved core processes present in both organisms. The signaling pathways found in the insect eye and in the vertebrate eye possess the same capacity for weak linkage inherited from the first eukaryotic single-celled organisms some two billion years ago. Behe sees the constraint in the particular designs, but not the deconstraint these designs provide. From a distance, a toy castle and a toy Eiffel Tower, both made of Lego blocks, look vastly different. Only on close inspection are the commonality and the clever interconvertibility of their component parts revealed.

Our third example, *Icons of Evolution*, critiques some of the older scientific evidence for evolution. Regarding that evidence, the author accuses "dogmatic Darwinists" of fraud. He omits citation of the most modern science, even though his examples largely concern phenotypic variation—a field that, we have shown, exploded in the last decade of the twentieth century. One of the icons, which he determines to have feet of clay, is the confusion of homology for common descent. He disputes that the anatomical similarities (the homology) of diverse vertebrate limbs (bats, porpoise, horse, and human) can be used as evidence that these vertebrates descended from a common ancestor with a limb. This argument was important to Darwin and all morphologists and paleontologists, so that demonstrating it to be fraudulent would be devastating to traditional and still-important anatomical evidence for evolution.[12]

Indeed, similarity alone is not enough to argue for descent from a common ancestor. (This point is thoroughly appreciated by scientists, so they are hardly naive in its application.) An octopus eye and a human eye are not homologous even though they look grossly similar, as we saw in Figure 38. As stated in *Icons of Evolution*, if we simply define "homology as common ancestry, how can we use homology as evidence for evolution?"[13]

Of course, the deeper we probe into the development, physiology, and anatomy of the octopus eye, the more we see how completely different it is from the vertebrate eye. But as we probe each vertebrate limb, the opposite happens. On the deepest molecular level of selector genes and signaling pathways in their development, the limbs of all vertebrates are strikingly similar. They form at exactly the same position relative to the segmented muscle blocks. The pectoral fins of fish use the same selector gene as the forelimbs of the mouse, whereas the pelvic fins use a different factor, the same as the hindlimbs. The detailed patterning of the front-to-back selector genes is the same in all limbs.

Where do the differences arise in the limbs of bats, porpoises, horses, and humans? As one might expect from facilitated variation, they come from the timing of, and amounts of, the secreted factors and selector genes affecting the growth of the various limb bones. The greatest differences are in the digits, reflecting their diversity. If we ask about the irreducible complexity of the limb, we see that it has avoided that problem in its evolution by the highly adaptive exploratory systems of the muscle, vascular system, and nerves—all of which migrate, proliferate, and make functional connections relative to the skeletal elements. The homology of limbs was one of the triumphs of evolutionary biology in the nineteenth century; they are more deeply understood than any other anatomical structure, and the modern molecular evidence for homology, its development, and its evolution, is unassailable.

Those motivated to question Darwin's theory found its weakest point in the origin of novelty and the sufficiency of random genetic change. This focus is quite reasonable, considering the vast amount of modern research that has given us our first real insights into the origin of the phenotype. The other two underpinnings of evolution theory, natural selection and heredity, were understood in outline even a hundred years ago; they have ascended in the past century to a high degree of mathematical and molecular sophistication.

Phenotypic variation, as a subject of molecular, cellular, and developmental inquiry, is a comparatively new field of study. The apparent weaknesses surrounding novelty and phenotypic variation have now been corrected. The questions raised by the creationists did not worry scientists, who in the past just accepted that they had incomplete and provisional answers.

Today's persuasive and consistent answers have come through molecular, cellular, and developmental experiments. On the sufficiency of random mutation, the evidence is strong that the phenotypic variation generated by the organism is not random but biased in a way related to the organism itself.

The explanation of how novelty is generated by facilitated variation, even if it did not exactly correspond to the questions the creationists asked, can now be seen as one of the strengths of a general theory of evolution. Largely because they derive from experiment, the mechanisms proposed for phenotypic variation have both great explanatory power and great verifiability. Molecular and cellular data can be added to the results of prior biologists and assembled into a coherent theory addressing the specific evolutionary problems of novelty and variation. These accomplishments should help open-minded people at all levels of scientific sophistication to realize that although many scientific questions remain in evolution and embryology, there are no apparent gaping holes in our theoretical understanding of evolution and no conspiracy of silence on scientific findings.

Meanwhile, Back on the Heath

We return to the doubts and concerns of people who understandably marvel at life and want to stand humbly on our planet. Does facilitated variation offer anything to people like the Reverend William Paley, who opened our narrative pondering the origin of the living world? His walk on the heath in 1802 was interrupted when he stumbled on a brass watch, an encounter that unleashed an introspective meditation about inferring the designer from the design: a human craftsman for the watch, the Supreme Creator for living beings. Paley's descendant, living in the twenty-first century, returned to the same heath but brought with her an education in modern biology. The younger Paley would want to tell her distinguished ancestor that his reservations about attempts to explain the creation of complex living things by natural causes, or anything less than a Divine Creator, have continued to resonate for a remarkably long time.

She would say that Darwin's theory of evolution, promulgated half a century after the elder Paley's death, was an ingenious attempt to reach a natural explanation of the origin of living things. For years, many doubted that it was true. Over time, two parts of a theory of evolution were proven beyond a doubt, the theory of heredity and the theory of natural selection. Ms. Paley might compare the questioning of the theory of heredity in our time to insisting that the world is flat. Similarly, she might affirm that natural selection is self-evident to many scientists and laypeople, especially to animal and plant breeders who regularly apply artificial selection to increase crop yields and manipulate animal breeds.

Remarkably, though, after almost two-hundred years, she could tell the elder Paley that his doubts about the origin of complex structural and functional organization and design in organisms can still inspire heated debate. The theory of phenotypic variability was obviously the weakest link in Darwin's general theory of evolution. The means for generating the amount and quality of phenotypic novelty were hardest for scientists and the lay public to comprehend, even when the requirement for random mutation was well established. She could now assure him that plausible answers have been found, although she understands that in his time it was impossible for him to imagine an alternative to divine creation. He would have needed to imagine what it has taken two centuries and tens of thousands of scientists to bring to light.

The younger Paley, perhaps having just finished this book, could report to her ancestor that the research results are not infinitely complicated, though full of abstruse details. Each organism has not evolved by its own rules. Variation has not percolated through all features of the phenotype. General principles of generating variation are evident and have been incorporated into a general understanding, a theory of facilitated variation. The theory may be still in progress, to be improved as more mechanisms unfold, but it is no longer accurate to say that science cannot explain the generation of novelty and the pace of evolutionary change. Nor is it correct to say that the greater the complexity of the organism, the harder it is to explain its evolution. Just the opposite. The special nature of the complexity is at the heart of the capacity to generate variation. Ironically, this complexity is dominated by conservation. Novelty itself has been deflated. Variation is facilitated largely because so much novelty is available in what is already possessed by the organism.

The question of faith remains, though, and the elder Paley might well return to it. Here, the younger Paley might advise her ancestor that she respects his efforts to deduce the Designer from the Design of living organisms, and to demarcate a clear line between human knowledge and faith. She offers her hope that, in light of two centuries of discoveries, he would draw his line between faith and science at a different place, one more defensible in light of the modern understandings. Finally, before they separated, she would reveal to her ancestor one great and simple truth. "You could never have imagined it," she might say, "but when we finally discovered how life was constructed, it turned out to be nothing like a brass watch or a divine creation beyond human comprehension. The secret lay in understanding the organism on its own terms."

Glossary

Adaptability, somatic The responses made by an organism to environmental change, which offset detrimental effects of the environment or that increase the organism's performance under adverse conditions.

Adaptation, evolutionary Heritable change in the phenotype of an organism that has been selected for the increased reproductive fitness it confers in a particular environment.

Adaptive cell behavior Refers to the fact that the core processes of cells respond to local environments and to cell-cell signals to adjust their outputs. When the organism evolves, these processes change combinations and amounts to generate the observable phenotype.

Allostery Refers to the fact that some proteins have two kinds of sites on the surface, one at which a function occurs and the other at which regulation of the function occurs. More deeply, it refers to the fact that allosteric proteins have not one but two conformations, or states, that differ in the degree of activity of the functional site and the regulatory site. Regulatory signals that stabilize the conformation with greater function are activators of function. Signals that stabilize the less functional conformation are inhibitors of function.

Arthropods Members of a phylum of animals having a body plan marked by an exoskeleton, jointed appendages, body segments, a ventral nerve cord, and no tail. Members include insects, crustaceans, spiders, centipedes, and trilobites (extinct).

Assimilation (*see also* Genetic assimilation) The heritable stabilization of a somatic adaptation by genetic change. It may occur through new mutation or by a resorting of the existing genetic variation in the population.

Baldwin effect James Mark Baldwin's proposal in 1896 of a path by which an organism might produce heritable phenotypic variation of benefit in the selective environment. He argued that when animals face an altered environment, they respond to it with a somatic adaptation enabling them to tolerate the environmental change. They are still under stress, even though they are viable enough to reproduce at least minimally. In subsequent generations, heritable changes arise in a few members of the population; these improve or modify the adaptation and increase the animal's reproductive fitness. By this scheme, genetic change does not have to precede phenotypic change; mutation follows during selection to improve on the change.

Body plan A global body organization comprising certain aspects of anatomy and a map of the compartments of signaling proteins and of selector proteins from expressed selector genes. Each animal phylum is distinguished by a unique body plan, and almost all have been conserved since the Cambrian period. It is first formed midway in embryonic development.

Cell types The kinds of differentiated cells of a multicellular organism. They share the same genotype but differ in the genes they express, the messenger RNAs and proteins they contain, and the cellular functions they perform.

Chordates Members of a phylum of animals including the vertebrates, the cephalochordates (lancelets), and the urochordates (ascidians, tunicates). All have a notochord, dorsal hollow nerve cord, gill slits (or branchial arches), and a tail extending beyond the anus.

***Cis*-regulation of genes** Refers to DNA sequences that are adjacent to a gene by which the gene's transcription is controlled through the binding of various transcription factors.

***Cis*-regulatory model of evolution** In the extreme model, phenotypic change in multicellular organisms that involves mostly the change of *cis*-

regulatory regions of genes, by which the time, place, and level of expression of the gene changes but the coding sequence does not.

Common ancestor An ancestor from which two or more lineages of organisms have evolved.

Compartment A region of the embryo in which one or a few selector genes are uniquely expressed, and in which one or a few signaling proteins are produced. By a middle stage of development, the animal embryo forms a map of these compartments.

Compartment map The spatial array of compartments in an animal. The map serves as a scaffold or platform for locating and building complex anatomical structures. Each phylum of animals has a distinctive map. The map is more conserved than are the kinds of anatomy and physiology built upon it.

Compartmentation The capacity to operate different conserved core processes at different places in the organism, and in fact to create those places.

Conservation The retention by a lineage of organisms of particular gene and protein sequences over long periods. The sequences undergo mutational changes, like all sequences, but most changes damage protein function. They are lethal and eliminated; the unchanged sequence is retained.

Conserved core processes The processes that generate the anatomy, physiology, and behavior of the organism in the course of its development (several hundred in number) and comprise the organism's phenotype. The various traits are generated by different combinations of the processes operating in different parts of their adaptive ranges of performance. Some of these processes have been unchanged (conserved) for hundreds of millions or even billions of years.

Constraint The term used to indicate that an organism cannot possess a particular kind of heritable phenotypic variation because it is lethal.

Convergence, evolutionary The term used when two organisms have similar structures performing similar functions, but they have evolved the structures independently.

Cytoplasm In a eukaryotic cell, the fluid space of the cell between the plasma membrane surface and the nucleus inside. In a prokaryotic cell, it is the fluid space between the plasma membrane surface and DNA strands inside.

Cytoskeleton Found extensively in eukaryotic but not prokaryotic cells; composed of extensive arrays of protein filaments that give shape to the cell and provide a scaffold on which materials are moved directionally.

Darwinian theory of evolution Proposed by Charles Darwin in 1859 to explain the origin of the diverse forms of organisms on earth by ongoing descent with modification from ancestors, rather than by a simultaneous creation and fixation of forms at the beginning of time. In Darwin's view, rephrased in modern terms, organisms within populations vary genetically and consequently in traits that affect their capacity to reproduce under the conditions at hand. In competition with one another and facing other pressures in the environment, the more fit organisms flourish and the less fit fail; this selection leaves a better-adapted subset. The replacement population is said to have evolved under selection, based on its genetic variation.

Deconstraint A counterpart to constraint; gained as a trade-off for constraint of the workings of the core processes. We argue that the core processes and their components are built in ways that lower constraint on the evolution of new regulatory connections between and within processes.

Derepression The activation of a gene by inhibiting its repressor. Usually refers to the process whereby small metabolites antagonize protein repressors that are bound to DNA, where they prevent transcription.

Descent with modification Darwin's term to denote that as organisms go though generations of descent from ancestors, they accumulate changes in their anatomy, physiology, and behavior. By implication, the changes are modifications of constituents and processes that were previously present, rather than entirely new ones.

Design Used here to mean a structure as it is related to function, not necessarily implying a designer.

Developmental biology The study not only of the embryonic development of organisms (usually plants, fungi, or animals) but also of their development through the larval, juvenile, and adult stages.

Diploid The state of having two sets of chromosomes. In higher organisms one set is contributed by the father and the other by the mother at fertilization.

DNA base sequence (*see also* Genome, Genotype) The ordering of the four constituents A, T, G, and C of the DNA chain, which is 3 billion bases long in the case of humans and about 140 million bases long in the case of the fruit fly.

Embryonic induction Change in the development of a multipotential region of the embryo in response to signals (inducers) from another region.

Engrailed A selector gene, originally discovered in fruit flies, encoding a transcription factor, the Engrailed protein. In insects the gene is expressed in the posterior compartment of each body segment. Its expression makes the development of the posterior compartment different from the anterior, because the engrailed protein turns on or off many other genes in the posterior compartment. Without the engrailed gene, the posterior compartment develops like the anterior.

Enzyme activity Refers to the kind of chemical reaction, and the degree to which it is accelerated, by proteins that are catalysts (substances that are not consumed in the reaction but increase its rate).

Enzyme induction An increased rate of synthesis of the RNA coding for an enzyme in response to exposure to an inducer.

Eukaryotes These include single-celled protists and multicellular plants, fungi, and animals. All eukaryotic cells possess internal membrane-bounded organelles such as the nucleus, secretory vesicles, mitochondria (or remnants), and chloroplasts (in plants). All eukaryotes have linear DNA pieces contained in the chromosomes. All have a cytoskeleton for cell shape, cell organization, and directional movement of materials in the cell.

Evolution (*see also* Darwinian theory of evolution) The descent of species from a common origin by a process of heritable phenotypic change; as opposed to special creation.

Evolvability Broadly, the capacity to evolve; includes both a variation component and selection component. Here we emphasize the variation component—that is, the capacity of the organism to generate viable phenotypic variation in response to genotypic variation, especially viable variation bestowing increased fitness.

Exon In eukaryotes, a segment of a gene, the RNA copy of which is spliced out and incorporated into the messenger RNA. All coding sequences in proteins are encoded in exons.

Exploratory behavior An adaptive behavior of certain cellular and developmental core processes, wherein they generate many, if not an unlimited number of, specific outcome states, any of which can be stabilized selectively by other kinds of agents. Examples include microtubules contacting chromosomes at mitosis; nerve axons contacting distant target cells or organs, and even ants searching for food.

Extracellular matrix In animals, a layer of insoluble materials (proteins) secreted by cells and deposited between them, to which cells attach and on which cells move.

Facilitated genotypic variation An organism's hypothetical response to stressful environmental conditions by making directed changes in DNA sequences that result in phenotypic changes of benefit to its survival against that stress. (Evidence for this kind of variation has been often sought but never found.)

Facilitated variation An explanation of the organism's generation of complex phenotypic change from a small number of random changes of the genotype. We posit that the conserved components greatly facilitate evolutionary change by reducing the amount of genetic change required to generate phenotypic novelty, principally through their reuse in new combinations and in different parts of their adaptive ranges of performance.

Fitness The capacity to contribute progeny to future generations.

Gastrulation A developmental process by which many cells on the surface of the embryo move inside and transform the original organization of the egg into that of the larva, juvenile, or adult. By the completion of gastrulation and neurulation (by which the nervous system is internalized), the body plan of the animal is formed.

Gene expression All the steps by which the DNA base sequence of a gene is converted to an active protein or functional RNA. In eukaryotic cells, the steps include transcription of the DNA sequence into an RNA copy, splicing and trimming of the RNA into a messenger RNA, translation of the messenger RNA into a polypeptide chain, folding of the chain into a protein, and activation of the protein.

Genetic assimilation A term invented by Conrad Waddington for the heritable stabilization of a somatic response. The heritable changes give improved fitness and are selected. The somatic response may or may not be adaptive for the condition of the environmental stimulus. At the end of the assimilation, the somatic response can be produced by the organism even in the absence of the environmental stimulus.

Genetic reassortment Genotypic variation resulting from genetic recombination and chromosome reassortment during sexual reproduction. The term often refers to changes of DNA base sequence (old mutations) that have been in the population for a time, not mutations that have arisen in the lifetime of the current individual.

Genetic variation, random (*see also* Variation, heritable phenotypic) Change in the sequence of DNA due to mutation, recombination, assortment of different chromosomes, and insertion of DNA from viruses and other organisms. The specific variation is unlinked to the environment and independent of selection.

Genome The entire DNA sequence of an organism (the ordering of the four elements A, T, G, and C). Sometimes also called *genotype*, but *genome* is a more general term referring to the base sequence of a species, such as the human genome, without referring to individuals of the species.

Genotype The entire DNA base sequence of an individual organism. Members of the population may differ slightly in their gene sequences, owing to genetic variation in the population.

Germ line A special group of cells of a multicellular organism that produce eggs or sperm; somatic cells, on the other hand, are not capable of inheritance.

Haploid The condition of having one set of chromosomes. The chromosome complement of the sperm and the egg after meiosis. In asexual organisms such as bacteria, haploidy is usually the normal state.

Heredity The property of a living organism whereby information about the specific character of the organism is passed to the next generation with high fidelity.

Homeosis Replacement of a missing part of an animal by another part, thus present twice. An example is the bithorax mutant of *Drosophila*, in which four wings rather than two are present and the balancing organs are absent.

Hox genes Selector genes that regulate compartment identity and are expressed in the posterior head and trunk of many animals. They are generally clustered together on a chromosome.

Hypoxia The condition of insufficient oxygen. Animals have many physiological responses to hypoxia, directed to increasing oxygen availability. They may produce substances that promote more oxygen release from hemoglobin or more blood vessel growth, or they may breathe more rapidly and increase lung capacity and heart rate.

Inducer A word with two meanings in biology. First, with reference to bacteria, an inducer is a small molecule, usually a foodstuff, that provokes the bacterium to make an enzyme protein of the kind to degrade the foodstuff for growth. Second, with reference to developing embryos, an inducer is a signal protein released by some cells that provokes nearby cells to develop in a way they would not have done if the inducer were absent. Both kinds of inductions are found to be permissive; that is, the cells are highly prepared to respond in a certain way and the inducer is but a trigger.

Instructive and permissive signaling Permissive signaling denotes that a complete response has been built into the receiver and then internally repressed. When the signal relieves the repression, the receiver unleashes its ready-made response. In instructive signaling, the response is not built in ahead of time and the signal must provide information for generating the response.

Interchangeability of cues Organisms often possess carefully poised switches to control either-or decisions in development and physiology. These can be thrown either by environmental cues such as temperature or by internal components encoded by genes. Often closely related organisms have interchanged environmental and genetic controls.

Intron In eukaryotes, a segment of a gene, the RNA copy of which is spliced out, discarded, and not incorporated into the messenger RNA. Introns may contain sequences important for regulating genes.

Isotropic variation Heritable phenotypic variation that is not directed toward the adaptive needs of the organism. Some biologists believe that if variation is really isotropic, selection must be the creative force shaping all adaptations.

Kinase An enzyme that adds a phosphate group to target proteins for the purpose of regulating their activity or other functions.

Lamarckian inheritance of acquired characteristics Lamarck's 1809 theory that when animals adapt physiologically or behaviorally to a stressful condition, they pass that adaptation to their offspring, who are then better adapted than the parents.

Lethal mutation Genetic changes that lead to loss of viability of the organism, often because of the loss of function of a component of a conserved core process.

Macromutation A large evolutionary change in a single step. Hugo de Vries thought he had found such changes in the evening primrose, which produced new species abruptly, but later work showed this plant to have an unusual hybrid instability. Macromutation is the opposite of micromutation, small evolutionary changes, of which many would be needed in succession to gain a large change. Darwin's view of evolution entailed micromutational change.

Master regulatory genes Genes that continuously express one or more transcription factors in certain differentiating cells. These master regulators, like selector proteins of a body plan compartment, activate or repress many target genes, determining the profile of RNAs and proteins of that cell type. Master regulatory genes are known for muscle, nerve, and fat.

Messenger RNA In eukaryotic cells, the kind of RNA copied in the nucleus from gene sequences encoding protein sequence information. The messenger RNA, which then bears the sequence information for the protein, carries the information to the cytoplasm, like a messenger; there it is translated to make the protein.

Metazoa The scientific term for the animal kingdom. Metazoa are all multicellular and require ready-made, complex foodstuffs. Their cells lack rigid walls. The 30 phyla of metazoa differ greatly in their anatomy and physiology and range from sponges to insects, to oysters, to humans.

Microtubule A stiff hollow protein filament in the cytoskeleton; the main structural element in nerve axons, the mitotic spindle, cilia, and flagella. Its rapid assembly and disassembly in the cell allows the population of microtubules to explore space and to be stabilized in different arrangements.

Mitochondrion The energy-producing organelle of a eukaryotic cell. It originated from a prokaryotic cell engulfed by an early eukaryotic cell ancestor.

Modern Synthesis (*see also* Neo-Darwinian theory) A version of Darwin's theory of variation and selection, consolidated in the 1940s with Mendelian theory. The concept of evolutionary adaptation was made paramount. Natural selection took center stage and the idea of inheritance of somatic adaptations (acquired characteristics) was purged as an explanation for the generation of phenotypic variation.

Modularity A plan of organization built around semi-independent units of integrated design.

Module One of several units in a larger design. The units may be complex and strongly integrated within themselves, but are more loosely linked in the larger plan so that individual units can be readily substituted without jeopardizing the larger structure.

Morphosis Changes in the organism's phenotype under stressful conditions, which do not help the organism adapt to the stress. The phenotypic variation may be fortuitously adaptive for a selective condition other than the one that provokes it, in which case it may be stabilized by heritable genetic variation.

Mutant An organism that has undergone a genetic change, usually one that has observable consequences.

Mutation (*see also* Random mutation) A change of the sequence of A, T, G, and C elements in DNA, due to causes such as chemical or radiation damage to the DNA, unrepaired errors in the replication of DNA, errors of recombination of DNA strands, movement of virus-like DNA sequences to new sites, and insertion or deletion of pieces of DNA. These changes occur randomly in the DNA sequence.

Natural selection Darwin's concept of the environment's effect on a population in which organisms differ in heritable phenotypic traits. Some individuals are eliminated as less fit for reproducing in that environment.

Neo-Darwinian theory A merger of Darwin's theory of natural selection, strengthened by August Weismann's strong negation of inheritance of acquired characters, with a gradualist view of evolutionary history and incorporating Mendelian inheritance and population genetics.

Neural crest cell During the embryonic development of vertebrates, neural crest cells arise at the edge of the central nervous system and migrate through the rest of the body, settling at various sites where they proliferate and differentiate. They have many options for differentiation, such as bone, cartilage, nerves, gland cells, pigment, or components of the heart. The particular differentiation depends on the signals at the site of settlement and the kind of neural crest cell, related to the site from which it came.

Norm of reaction The range of various phenotypes expressed when an organism reacts to a range of environmental conditions such as temperature, humidity, crowding, or kind of food. Some responses confer adaptive benefit; others (called morphoses) are nonadaptive responses to environmental stress. The norm of reaction represents all the phenotypic variation an organism can produce without changing the genotype.

Novelty (*see* Phenotypic novelty)

Pangenesis Darwin's theory, later refuted, of the inheritance of acquired characteristics. He proposed that cells of the body produce informational particles in amounts related to their physiological use, and that the particles collect in germ cells, the egg and sperm precursors in the gonads. Then, in the embryo developing from those germ cells, the particles direct greater or lesser development of parts according to the parents' usage.

Permissive signaling (*see* Instructive and permissive signaling)

Phenotype Includes all the observable and functional features (traits) of an organism, that is, its anatomy, physiology, development, and behavior, and also all its conserved core processes. Some aspects of phenotype are heritable and some are dictated by the environment.

Phenotypic novelty Large complex changes of phenotype such as the eye, the hand, the beak, or the evolution of humans from bacteria. The origin of novelty is perhaps the greatest unanswered question in evolution.

Phenotypic variation (*see also* **Darwinian theory of evolution; Variation)** The differences in phenotype of members of a population of organisms. (It can be generally observed that no two members are exactly the same.) Some differences are heritable and passed to offspring. Other differences are nonheritable adaptations to the environment and change when the environment changes.

Phylogeny A branching tree of descent of organisms from ancestors.

Phylotypic stage A middle stage of development when the body plan of a phylum is first present; includes the map of compartments of expression of selector genes and secreted signaling proteins. Embryos of the different classes of a phylum of animals look most alike at this stage, before the specialized organs and cell types are developed at particular sites selected by the compartments. In chordates, the phylotypic stage manifests a newly developed dorsal hollow nerve cord, gill slits, the beginnings of a tail, and the notochord, as well as the compartment map.

Phylum A group of animals sharing a body plan, defined in the past as a unique suite of anatomical traits but now including a unique map of compartments of selector proteins and signaling proteins.

Physiological adaptability One kind of somatic adaptability; involves physiological as opposed to developmental responses to environmental changes. The responses are usually reversible in minutes to weeks when the stimulus is withdrawn.

Physiological variation Differences in the responses of individual animals to environmental challenges. Attributable to different exposures and to differing responsiveness, which in turn may reflect prior exposures, genetic differences of the individuals, or both.

Plasticity The capacity of organisms with the same genotype to vary in phenotype, according to varying environmental conditions. Or the capacity for alteration of the neural circuits and synapses of the nervous system in response to experience or injury, by forming new circuits and synapses and eliminating old ones. More generally, the somatic adaptability of the organism in response to environmental or genetic change.

Pleiotropy Occurs when a genetic change has conflicting effects in different regions of the embryo or at different times in the life cycle; that is, when a change for the better occurs in one place, a change for the worse in another. Compartments of expressed selector genes mitigate these effects by causing different changes of a target gene's expression at different sites in the embryo, rather than one kind of change serving in multiple locations.

Polyphenism Referring to animals, the condition of having two or more phenotypes, or phenes, that can be developed from the same genotype, alternatively or sequentially. The larva and adult are sequential phenes for many animals. The queen and worker are alternative phenes for honeybees.

Prokaryotes These include the smallest free-living organisms, the eubacteria and archaebacteria (or archaea). All lack a nucleus and internal membrane-bounded organelles. All have a genome of circular DNA contained in the cytoplasm. They divide asexually, some as rapidly as every 20 minutes. Their processes for making DNA, RNA, protein, usable energy, and cell constituents are basically the same as those of eukaryotic cells and multicellular organisms.

Protist A eukaryotic single-celled organism, such as an amoeba or paramecium. The first protists may have arisen two billion years ago. The diversity of protists is enormous. They used to be called protozoa, but it is now recognized that they are less related to animals in their ancestry than are plants and fungi.

Random mutation Changes of the DNA base sequence (the order of A, T, G, and C) that are not directed to particular regions of the genome by the selective conditions of the environment. Some biologists reserve the word *mutation* for changes that have occurred in the lifetime of the individual under consideration, whereas all older mutations, traded around by recombination and chromosome reassortment in the course of sexual reproduction, are known as genetic variation.

Recombination Rearrangement of the DNA to produce a new overall sequence by any of several means including crossover between sister chromatids at meiosis, translocation mediated by viruses, and breakage and rejoining of chromosomes. Reassortment usually refers to the random choice of which maternal and paternal chromosomes are retained in the germ cell after meiosis, although it may refer to genetic variation generated by a variety of means.

Regulatory mutation Genetic changes in either the coding or noncoding region of genes that serve to bring together core processes in different combinations at different times and places. They can also select on the processes for different parts of their adaptive ranges of performance. Included are changes of regulatory DNA (*see Cis*-regulation of genes) as well as changes of splicing, translational control, protein activation, and protein destruction.

Replication The synthesis of new DNA from old DNA, as an exact copy. During the copying, replication errors occur.

Robustness The resistance of the phenotype (anatomy, physiology, or behavior) to environmental or genetic change.

Selection, *see* Natural selection

Selector genes Genes, encoding transcription factors, that are expressed in the compartments of an animal's body plan and that serve to distinguish each compartment. The selector proteins of each compartment activate or repress a suite of target genes for conserved core processes selected to occur or not occur in the compartment, and they activate their own continued expression. Hox genes, for example, are selector genes.

Signal transduction The process of receiving a signal at the cell surface and relaying it through the cytoplasm by setting off a series of controlled chemical changes internally. In the end the cell's particular response is triggered.

Somatic adaptability (*see* Adaptability)

Somatic cells The cells of the body that experience stresses of the environment and respond to them by somatic adaptation. They include the muscles, nerves, bones, and skin, all the differentiated cells resulting from embryonic development. They do not include germ cells, which are the only cells able to transmit genes to the next generation.

Stabilizing selection (see Genetic assimilation) Occurrence of mutations when the environment is stable or when it fluctuates around a norm; the mutations reduce the sources of variation in the phenotype and produce more stability.

Substrate A molecule that undergoes a chemical transformation on the surface of an enzyme protein. Most small molecules within the cell are substrates of different enzymes as the cell's complex metabolism is accomplished.

Target genes Those genes activated and repressed by selector transcription factors (encoded by selector genes) or master regulatory factors (encoded by master regulatory genes).

Transcription The conserved core process by which a DNA base sequence is copied into RNA. In eukaryotic cells, the RNA is cut to remove intron sequences and spliced to make messenger RNA.

Transcription factor A protein that binds to the DNA regulatory region of a gene and increases or decreases the expression of that gene.

Translation The synthesis of a protein molecule consisting of hundreds of amino acid units connected in a unique sequence. Messenger RNA provides the sequence information, and the ribosome associated with that RNA connects the amino acids.

Variants Members of a population of organisms that differ in their genetic composition (genetic variants) or in their phenotype (phenotypic variants).

Variation, heritable phenotypic (*see also* Genetic variation, Isotropic variation, Phenotypic variation) Darwin's theory that heritable phenotypic variants inevitably arise in populations of organisms, and that those that are more fit for reproducing in the environmental conditions at hand eventually overtake the population.

Weak regulatory linkage A form of regulation easily devised and easily changed to link core processes together in new combinations or to select one part of their adaptive range of performance. The regulatory signal provides little information about the outcome, whereas the receiver is maximally informed. Thus, many core processes are built to have two states of operation, on or off, and to be receptive to signals. The signal merely selects one of the states by stabilizing it. The signals are said to work by permissive rather than instructive interactions.

Notes

INTRODUCTION
A Clock on the Heath

1. W. Paley, *Paley's natural theology* (London: C. Knight, 1836), 1.

2. Ibid., 55.

3. C. Darwin, *On the origin of species by means of natural selection, or the preservation of favoured races in the struggle for life* (London: John Murray, 1859); S. B. Carroll, J. K. Grenier, and S. D. Weatherbee, *From DNA to diversity: Molecular genetics and the evolution of animal design* (Malden, Mass: Blackwell Science, 2001); L. B. Radinsky, *The evolution of vertebrate design* (Chicago: University of Chicago Press, 1987).

4. D. Sobel, *Longitude: The true story of a lone genius who solved the greatest scientific problem of his time* (New York: Walker, 1995).

5. S. Panda, J. B. Hogenesch, and S. A. Kay, "Circadian rhythms from flies to human," *Nature* 417 (2002): 329.

ONE
The Sources of Variation

1. J. B. Lamarck, "Zoological philosophy: An exposition with regard to the natural history of animals" in A. S. Packard, *Lamarck, the founder of evolution* (New York: Longmans, Green, 1901), 316.

2. Ibid., 351.

3. J. B. Lamarck, *Zoological philosophy: An exposition with regard to the natural history of animals* (Chicago: University of Chicago Press, 1984), 127.

4. C. Darwin, *The variation of animals and plants under domestication*, vol. 2 (London: John Murray, 1868), 367.

5. A. Weismann, *The germ-plasm: A theory of heredity* (New York: Scribner, 1893), 5.

6. W. Bateson, *Materials for the study of variation treated with especial regard to discontinuity in the origin of species* (London: Macmillan, 1894), 6.

7. A. H. Sturtevant, "Thomas Hunt Morgan, September 25, 1866–Decem-

ber 4, 1945," in *National Academy of Sciences (USA), biographical memoirs* (New York: National Academy of Sciences, 1959), 283.

8. I. Shine and S. Wrobel, *Thomas Hunt Morgan, pioneer of genetics* (Lexington: University Press of Kentucky, 1976).

9. J. Cairns, J. Overbaugh, and S. Miller, "The origin of mutants," *Nature* 335 (1988): 142.

10. S. J. Gould, *The structure of evolutionary theory* (Cambridge, Mass.: Belknap Press of Harvard University Press, 2002), 503; J. Huxley, *Evolution: The modern synthesis* (London: Allen and Unwin, 1942); T. G. Dobzhansky, *Genetics and the origin of species* (New York: Columbia University Press, 1982).

11. Gould, *Structure of evolutionary theory*; A. C. Milner, *Dino-birds: From dinosaurs to birds* (London: Natural History Museum, 2002).

12. V. Hamburger, in *The evolutionary synthesis: Perspectives on the unification of biology*, ed. E. Mayr and W. B. Provine (Cambridge, Mass.: Harvard University Press, 1980), 96.

13. Gould, *Structure of evolutionary theory*, 60.

14. S. Wright, "Evolution in Mendelian populations," *Genetics* 16 (1931): 147.

TWO
Conserved Cells, Divergent Organisms

1. M. Spencer et al., "Analyzing the order of items in manuscripts of *The Canterbury Tales*," *Computers and the Humanities* 37 (2003): 97.

2. N. Eldredge and S. J. Gould, "Punctuated equilibria: An alternative to phyletic gradualism," in *Models in Paleobiology*, ed. T. J. M. Schopf (San Francisco: Freeman, Cooper, 1972), 82.

3. J. M. Peregrin-Alvarez, S. Tsoka, and C. A. Ouzounis, "The phylogenetic extent of metabolic enzymes and pathways," *Genome Research* 13 (2003): 422.

4. F. Crick, *Life itself: Its origin and nature* (New York: Simon and Schuster, 1981), 141.

5. C. R. Woese, "Bacterial evolution," *Microbiological Review* 51 (1987): 221.

6. D. E. Canfield and A. Teske, "Late Proterozoic rise in atmospheric oxygen concentration inferred from phylogenetic and sulphur-isotope studies," *Nature* 382 (1996): 127.

7. J. W. Valentine and D. Jablonski, "Morphological and developmental macroevolution: A paleontological perspective," *International Journal of Developmental Biology* 47 (2003): 517.

8. A. Adoutte et al., "The new animal phylogeny: Reliability and implications," *Proceedings of the National Academy of Sciences (USA)* 97 (2000): 4453; S. Conway-Morris, "The Cambrian 'explosion' of metazoans and molecular biology: Would Darwin be satisfied?" *International Journal of Developmental Biology* 47 (2003): 505.

9. Eldredge and Gould, "Punctuated equilibria."

10. N. H. Shubin, "Origin of evolutionary novelty: Examples from limbs," *Journal of Morphology* 252 (2002): 15.

11. J. G. Kingsolver and M. A. R. Koehl, "Selective factors in the evolution of insect wings," *Annual Reviews of Entomology* 39 (1994): 425.

THREE

Physiological Adaptability and Evolution

1. D. J. Depew and B. H. Weber, *Darwinism evolving: Systems dynamics and the genealogy of natural selection* (Cambridge, Mass.: MIT Press, 1995); M. J. West-Eberhard, *Developmental plasticity and evolution* (Oxford: Oxford University Press, 2003), 535.

2. West-Eberhard, *Developmental plasticity*, 116.

3. J. A. M. Baldwin, "A new factor in evolution," *American Naturalist* 30 (1896): 441; H. F. Osborn, *A mode of evolution requiring neither natural selection nor the inheritance of acquired characters (organic selection)* (New York: New York Academy of Science, 1896); C. L. Morgan, *Habit and instinct* (London: E. Arnold, 1896).

4. I. I. Schmalhausen, *Factors in evolution: The theory of stabilizing selection*, ed. T. Dobzhansky (Chicago: University of Chicago Press, 1986).

5. C. H. Waddington, "Genetic assimilation of an acquired character," *Evolution* 7 (1953): 118.

6. S. L. Rutherford and S. Lindquist, "Hsp90 as a capacitor for morphological evolution," *Nature* 396 (1998): 336.

7. C. Queitsch, T. A. Sangster, and S. Lindquist, "Hsp90 as a capacitor of phenotypic variation," *Nature* 417 (2002): 618.

8. West-Eberhard, *Developmental plasticity*, 151; C. D. Schlichting and M. Pigliucci, *Phenotypic evolution: A reaction norm perspective* (Sunderland, Mass.: Sinauer, 1998), 315.

9. G. G. Simpson, "The Baldwin effect," *Evolution* 7 (1953): 115; Schlichting and Pigliucci, *Phenotypic evolution*, 315.

10. H. F. Nijhout, "When developmental pathways diverge," *Proceedings of the National Academy of Sciences (USA)* 96 (1999): 5348.

11. A. Meyer, "Morphometrics and allometry in the tropically polymorphic cichlid fish, *Cichlasoma citrinellum*: Alternative adaptations and ontogenetic changes in shape," *Journal of Zoology (London)* 221 (1990): 237.

12. A. Huysseune, "Phenotypic plasticity in the lower pharyngeal jaw dentition of *Astatoreochromis alluaudi* (Teleostei: Cichlidae)," *Archives of Oral Biology* 40 (1995): 1005.

13. West-Eberhard, *Developmental plasticity*; M. J. West-Eberhard, "Phenotypic plasticity and the origins of diversity," *Annual Review of Ecology and Systematics* 20 (1989): 249.

14. E. B. Wilson, *The cell in development and inheritance* (New York: Columbia University, 1900), 144; N. Stevens, "Studies in spermatogenesis II. A

comparative study of the heterochromososmes in certain species of Coleoptera, Hemiptera and Lepidoptera with especial reference to sex determination," *Carnegie Institution of Washington Publication* (1906): 1; E. B. Wilson, "Studies on chromosomes III. The sexual differences of the chromosome group in Hemiptera with some considerations of the determination and inheritance of sex," *Journal of Experimental Biology and Medicine* 3 (1906): 1. S. F. Gilbert, "The embryological origins of the gene theory," *Journal of the History of Biology* 11 (1978): 320.

15. P. S. Western et al., "Temperature-dependent sex determination in the American alligator: Expression of SF1, WT1 and DAX1 during gonadogenesis," *Gene* 241 (2000): 223.

16. D. Crews, "Sex determination: Where environment and genetics meet," *Evolution and Development* 5 (2003): 50.

17. K. Semsar and J. Godwin, "Social influences on the arginine vasotocin system are independent of gonads in a sex-changing fish," *Journal of Neuroscience* 23 (2003): 4386.

18. J. J. Emerson et al., "Extensive gene traffic on the mammalian X chromosome," *Science* 303 (2004): 537.

19. J. A. Graves, "From brain determination to testis determination: Evolution of the mammalian sex-determining gene," *Reproduction, Fertility, and Development* 13 (2001): 665.

20. W. Harvey, *The circulation of the blood and other writings* (London: J. M. Dent, 1990), 92; J. W. Severinghaus, "Fire-air and dephlogistication: Revisionisms of oxygen's discovery," *Advances in Experimental Medicine and Biology* 543 (2003): 7.

21. M. Nikinmaa, "Haemoglobin function in vertebrates: Evolutionary changes in cellular regulation in hypoxia," *Respiration Physiology* 128 (2001): R317.

22. R. M. Wells, "Evolution of haemoglobin function: Molecular adaptations to environment," *Clinical and Experimental Pharmacology and Physiology* 26 (1999): 591.

23. N. B. Terwilliger, "Functional adaptations of oxygen-transport proteins," *Journal of Experimental Biology* 201 (1998): 1085.

24. Nikinmaa, "Haemoglobin function."

25. West-Eberhard, *Developmental plasticity.*

FOUR
Weak Regulatory Linkage

1. J. C. Gerhart and M. W. Kirschner, *Cells, embryos, and evolution: Toward a cellular and developmental understanding of phenotypic variation and evolutionary adaptability* (Boston: Blackwell Science, 1997); M. Kirschner and J. Gerhart, "Evolvability (perspective)," *Proceedings of the National Academy of Sciences (USA)* 95 (1998): 8420.

2. A. Lwoff and A. Ullmann, eds., *Origins of molecular biology: A tribute to Jacques Monod* (New York: Academic Press, 1979).

3. F. Jacob, in ibid., 100.

4. U. Alon, "Biological networks: The tinkerer as an engineer," *Science* 301 (2003): 1866.

5. C. B. Harvey and G. R. Williams, "Mechanism of thyroid hormone action," *Thyroid* 12 (2002): 441.

6. J. Monod and F. Jacob, "Teleonomic mechanisms in cellular metabolism, growth, and differentiation," *Cold Spring Harbor Symposium on Quantitative Biology* 26 (1961): 389.

7. T. H. Morgan, *Embryology and genetics* (New York: Columbia University Press, 1934), 10.

8. H. Spemann, *Embryonic development and induction* (New Haven: Yale University Press, 1938).

9. *Oxford English Dictionary* (Oxford: Oxford University Press, 2000).

10. E. M. De Robertis et al., "Molecular mechanisms of cell-cell signaling by the Spemann-Mangold organizer," *International Journal of Developmental Biology* 45 (2001): 189.

11. A. Lwoff, in Lwoff and Ullmann, *Origins of molecular biology*, 14; J. Monod, J. Wyman, and J.-P. Changeux, "On the nature of allosteric transitions: A plausible model," *Journal of Molecular Biology* 12 (1965): 85.

12. R. U. Lemieux and U. Spohr, "How Emil Fischer was led to the lock and key concept for enzyme specificity," *Advances in Carbohydrate Chemistry and Biochemistry* 50 (1994): 1.

13. H. F. Judson, *The eighth day of creation: Makers of the revolution in biology* (Plainview, N.Y.: Cold Spring Harbor Laboratory Press, 1996), 579.

14. T. Pawson and P. Nash, "Assembly of cell regulatory systems through protein interaction domains," *Science* 300 (2003): 445.

15. M. Z. Ludwig, N. H. Patel, and M. Kreitman, "Functional analysis of eve stripe 2 enhancer evolution in *Drosophila*: Rules governing conservation and change," *Development* 125 (1998): 949.

FIVE
Exploratory Behavior

1. J. C. Gerhart and M. W. Kirschner, *Cells, embryos, and evolution: Toward a cellular and developmental understanding of phenotypic variation and evolutionary adaptability* (Boston: Blackwell Science, 1997), 146.

2. C. Darwin, *On the origin of species by means of natural selection, or the preservation of favoured races in the struggle for life* (London: John Murray, 1859), 186.

3. M. Kirschner and T. Mitchison, "Beyond self-assembly: From microtubules to morphogenesis," *Cell* 45 (1986): 329.

4. C. Detrain, J. L. Deneubourg, and J. M. Pasteels, eds., *Information processing in social insects* (Basel: Birkhäuser, 1999).

5. A. Chisholm and M. Tessier-Lavigne, "Conservation and divergence of axon guidance mechanisms," *Current Opinion in Neurobiology* 9 (1999): 603.

6. J. Yuan and H. R. Horvitz, "A first insight into the molecular mechanisms of apoptosis," *Cell* 116 (2004): S53.

7. V. Hamburger, "History of the discovery of neuronal death in embryos," *Journal of Neurobiology* 23 (1992): 1116; R. Levi-Montalcini et al., "Nerve growth factor: From neurotrophin to neurokine," *Trends in Neuroscience* 19 (1996): 514.

8. N. Kasthuri and J. W. Lichtman, "The role of neuronal identity in synaptic competition," *Nature* 424 (2003): 426.

9. B. L. Schlaggar and D. D. O'Leary, "Patterning of the barrel field in somatosensory cortex with implications for the specification of neocortical areas," *Perspectives in Developmental Neurobiology* 1 (1993): 81.

10. K. C. Catania, "Barrels, stripes, and fingerprints in the brain—implications for theories of cortical organization," *Journal of Neurocytology* 31 (2002): 347.

11. E. Foeller and D. E. Feldman, "Synaptic basis for developmental plasticity in somatosensory cortex," *Current Opinion in Neurobiology* 14 (2004): 89.

12. D. O. Hebb, *The organization of behavior: A neuropsychological theory* (New York: Wiley, 1949); Y. Goda and G. W. Davis, "Mechanisms of synapse assembly and disassembly," *Neuron* 40 (2003): 243.

13. N. Ferrara, H. P. Gerber, and J. LeCouter, "The biology of VEGF and its receptors," *Nature Medicine* 9 (2003): 669.

14. C. J. Schofield and P. J. Ratcliffe, "Oxygen sensing by HIF hydroxylases," *Nature Reviews: Molecular Cell Biology* 5 (2004): 343.

SIX
Invisible Anatomy

1. M. Kieny, A. Mauger, and P. Sengel, "Early regionalization of somitic mesoderm as studied by the development of axial skeleton of the chick embryo," *Developmental Biology* 28 (1972): 142.

2. H. Driesch (1892), "The potency of the first two cleavage cells in echinoderm development: Experimental production of partial and double formations," in *Foundations of experimental embryology*, ed. B. H. Willier and J. M. Oppenheimer, (New York: Hafner Press, 1974), 38.

3. H. Spemann, *Embryonic development and induction* (New Haven: Yale University Press, 1938).

4. L. Wolpert, "Positional information and the spatial pattern of cellular differentiation," *Journal of Theoretical Biology* 25 (1969): 1; idem, "Positional information and pattern formation," *Current Topics in Developmental Biology* 6 (1971): 183.

5. S. E. Fraser and R. M. Harland, "The molecular metamorphosis of experimental embryology," *Cell* 100 (2000): 42.

6. A. Garcia-Bellido, "The engrailed story," *Genetics* 148 (1998): 539.

7. A. Garcia-Bellido and P. Santamaria, "Developmental analysis of the wing disc in the mutant engrailed of *Drosophila melanogaster*," *Genetics* 72 (1972): 87.

8. E. B. Lewis, "A gene complex controlling segmentation in *Drosophila*," *Nature* 276 (1978): 565.

9. J. J. Stuart et al., "A deficiency of the homeotic complex of the beetle *Tribolium*," *Nature* 350 (1991): 72.

10. P. A. Lawrence, *The making of a fly: The genetics of animal design* (Oxford: Blackwell Scientific Publications, 1992).

11. J. M. Slack, P. W. Holland, and C. F. Graham, "The zootype and the phylotypic stage," *Nature* 361 (1993): 490.

12. D. Duboule and P. Dolle, "The structural and functional organization of the murine HOX gene family resembles that of *Drosophila* homeotic genes" *EMBO Journal* 8 (1989): 1497.

13. E. Boncinelli, M. Gulisano, and V. Broccoli, "Emx and Otx homeobox genes in the developing mouse brain," *Journal of Neurobiology* 24 (1993): 1356.

14. J. C. Gerhart and M. W. Kirschner, *Cells, embryos, and evolution: Toward a cellular and developmental understanding of phenotypic variation and evolutionary adaptability* (Boston: Blackwell Science, 1997).

15. C. J. Lowe et al., "Anteroposterior patterning in hemichordates and the origins of the chordate nervous system," *Cell* 113 (2003): 853.

16. G. Pangaiban et al., "The origin and evolution of animal appendages," *Proceedings of the National Academy of Sciences USA* 94 (1997): 5162.

17. Gerhart and Kirschner, *Cells, embryos, and evolution,* 379.

18. G. Von Dassow and G. M. Odell, "Design and constraints of the *Drosophila* segment polarity module: Robust spatial patterning emerges from intertwined cell state switches," *Journal of Experimental Zoology* 294 (2002): 179.

19. G. C. Williams, *Natural selection: Domains, levels, and challenges* (New York: Oxford University Press, 1992).

20. G. Ruvkun and J. Giusto, "The Caenorhabditis elegans heterochronic gene lin-14 encodes a nuclear protein that forms a temporal developmental switch," *Nature* 338 (1989): 313; N. A. Moran, "Adaptation and constraint in the complex life cycles of animals," *Annual Review of Ecology and Systematics* 25 (1994): 573.

SEVEN
Facilitated Variation

1. M. J. West-Eberhard, *Developmental plasticity and evolution* (Oxford: Oxford University Press, 2003); G. G. Simpson, "The Baldwin effect," *Evolution* 7 (1953): 115.

2. L. W. Ancel and W. Fontana, "Plasticity, evolvability, and modularity in RNA," *Journal of Experimental Zoology* 288 (2000): 242.

3. West-Eberhard, *Developmental plasticity*.

4. J. C. Gerhart and M. W. Kirschner, *Cells, embryos, and evolution: Toward a cellular and developmental understanding of phenotypic variation and evolutionary adaptability* (Boston: Blackwell Science, 1997), 580.

5. S. B. Carroll, J. K. Grenier, and S. D. Weatherbee, *From DNA to diversity: Molecular genetics and the evolution of animal design* (Malden, Mass: Blackwell Science, 2001).

6. J. W. Fondon and H. R. Garner, "Molecular origins of rapid and continuous morphological evolution," *Proceedings of the National Academy of Sciences (USA)* 101 (2004): 18058; M. Ronshaugen, N. McGinnis, and W. McGinnis, "Hox protein mutation and macroevolution of the insect body plan," *Nature* 415 (2002): 914.

7. P. R. Grant, *Ecology and evolution of Darwin's finches* (Princeton: Princeton University Press, 1986); J. Weiner, *The beak of the finch: A story of evolution in our time* (New York: Knopf, 1994).

8. I. I. Schmalhausen, *Factors in evolution: The theory of stabilizing selection*, ed. T. Dobzhansky (Chicago: University of Chicago Press, 1986); West-Eberhard, *Developmental plasticity*; C. H. Waddington, "Genetic assimilation of an acquired character," *Evolution* 7 (1953): 118; S. L. Rutherford and S. Lindquist, "Hsp90 as a capacitor for morphological evolution," *Nature* 396 (1998): 336.

9. A. Abzhanov et al., "Bmp4 and morphological variation of beaks in Darwin's finches," *Science* 305 (2004): 1462.

10. M. D. Shapiro et al., "Genetic and developmental basis of evolutionary pelvic reduction in three-spine sticklebacks," *Nature* 428 (2004): 717.

11. Fondon and Garner, "Molecular origins"; Ronshaugen, McGinnis, and McGinnis, "Hox protein mutation."

12. D. N. Reznick, M. Mateos, and M. S. Springer, "Independent origins and rapid evolution of the placenta in the fish genus *Poeciliopsis*," *Science* 298 (2002): 1018; M. Kirschner and J. Gerhart, "Evolvability (perspective)," *Proceedings of the National Academy of Sciences (USA)* 95 (1998): 8420.

13. P. Dehal et al., "The draft genome of *Ciona intestinalis*: Insights into chordate and vertebrate origins," *Science* 298 (2002): 2157.

14. H. L. True, I. Berlin, and S. L. Lindquist, "Epigenetic regulation of translation reveals hidden genetic variation to produce complex traits," *Nature* 431 (2004): 184.

EIGHT
The Plausibility of Life

1. D. J. Depew and B. H. Weber, *Darwinism evolving: Systems dynamics and the genealogy of natural selection* (Cambridge, Mass.: MIT Press, 1995), 485.

2. R. Dawkins, in *Artificial life: The proceedings of an interdisciplinary*

workshop on the synthesis and simulation of living systems, ed. C. G. Langton (Redwood City, Calif.: Addison-Wesley, 1989), 201; J. C. Gerhart and M. W. Kirschner, *Cells, embryos, and evolution: Toward a cellular and developmental understanding of phenotypic variation and evolutionary adaptability* (Boston: Blackwell Science, 1997), 580.

3. J. Lowe, F. Van Den Ent, and L. A. Amos, "Molecules of the bacterial cytoskeleton," *Annual Review of Biophysics and Biomolecular Structure* 33 (2004): 177.

4. B. Jungnickel, T. A. Rapoport, and E. Hartmann, "Protein translocation: Common themes from bacteria to man," *FEBS Letters* 346 (1994): 73.

5. R. F. Doolittle, "The origins and evolution of eukaryotic proteins," *Philosophical Transactions of the Royal Society of London. Series B* 349 (1995): 235.

6. H. Hegyi and P. Bork, "On the classification and evolution of protein modules," *Journal of Protein Chemistry* 16 (1997): 545.

7. H. Kitano, "Systems biology: A brief overview," *Science* 295 (2002): 1662; D. E. Goldberg, *Genetic algorithms in search, optimization, and machine learning* (Reading, Mass.: Addison-Wesley, 1989).

8. M. E. Csete and J. C. Doyle, "Reverse engineering of biological complexity," *Science* 295 (2002): 1664.

9. "Evolution," in *The Catholic encyclopedia*, ed. C. G. Herbermann et al. (New York: Appleton, 1909), 5:654.

10. P. E. Johnson, *Darwin on trial* (Downers Grove, Ill.: InterVarsity Press, 1993), 72–73.

11. M. J. Behe, *Darwin's black box: The biochemical challenge to evolution* (New York: Free Press, 1996), 22, 39.

12. J. Wells, *Icons of evolution: Science or myth?* (Washington, D.C.: Regnery Publishing, 2000), 244.

13. Ibid., 77.

Index